The Fermented Vegetables Manual

Enjoy Krauts & Pickles & Kimchis the Right Way to Improve Skin, Health, and Happiness

TRACY HUANG
COVER BY LARA IAKOVENKO

All rights reserved. No part of this publication may be reproduced, stored in a retrieval system, or transmitted, in any form or by any means – electronic, mechanical, photocopying, recording, or others – without the prior written permission of the author and the publishers.

Please note that the scanning, uploading, and distribution of this book via the Internet or via any other means without the permission of the author is illegal and will be punished by the law. So, please purchase only authorized electronic editions, and do not participate in or encourages electronic piracy of copyrighted materials.

A Note to the Reader:

The information provided in this book is designed to provide helpful information on the subjects discussed. This book is not meant to be used, nor should it be used, to diagnose or treat any medical condition. For diagnosis or treatment of any medical problem, consult your own physician. The publisher and author are not responsible for any specific health or allergy needs that may require medical supervision and are not liable for any damages or negative consequences from any treatment, action, application or preparation, to any person reading or following the information in this book. Any references included are provided for informational purposes only. Readers should be aware that any websites or links listed in this book may change.

*To my husband, Chris,
who supports and believes in whatever I do.*

The Fermented Vegetables Manual

CONTENTS

INTRODUCTION   1

SECTION I: THE BIG PICTURE

Chapter 1: Meet Your One Hundred Trillion Allies   9
Chapter 2: The Civil War and Peace Agreements   29
Chapter 3: Consuming Fermented Vegetables - A Piece of the Puzzle   47

SECTION II: MASTER VEGETABLE FERMENTATION IN SIX STEPS

Chapter 4: Step One – Demystify Your Top Concerns   65
Chapter 5: Step Two - Prepare Essential Items   89
Chapter 6: Step Three – Make It Happen!   107
Chapter 7: Step Four – Eat Fermented Vegetables the Right Way   127
Chapter 8: Step Five – Store Them Well   139
Chapter 9: Step Six – Go Beyond Knowledge   143

SECTION III: VEGETABLE FERMENTATION DISCLAIMER AND FINAL THOUGHTS

Chapter 10: Oops… Vegetable Fermentation Is Not the "Holy Grail"   149

Chapter 11: Final Thoughts   153

SECTION IV: EXPERIMENTS AND RECIPES

Chapter 12: Sixteen Vegetable Fermentation Experiments   159
Chapter 13: Incorporating Fermented Vegetables into Daily Life   183

METRIC CONVERSION TABLES   199

RESOURCES   201

END NOTES   209

ACKNOLWEDGEMENTS   231

ABOUT THE AUTHOR   235

INTRODUCTION

This is a book about fermenting vegetables; yet it is about more than vegetable fermentation. I compiled this book to hold your hand and help you go deeper into your own body; to explore the world of microbes. Microbes actually have 10 times more cells than human cells and perform life-sustaining activities that you might not even be aware of. This book also helps you understand the role that probiotic supplements and fermented foods can play in the bigger picture.

Currently, there are many books related to making fermented foods. Yet, when I asked around to do my research, I realized that there were many people also very interested in why they should ferment; a lot of them were scared of the word "fermentation", as they associated it with rotten and moldy foods. They had other health-related concerns that stop them from even igniting the thought of trying to ferment.

I discovered that there is a gap between people who are ignorant about fermentation and who are afraid that the fermenting process itself might generate health issues. The books currently put on the market that are focused on teaching how to ferment are marketed to people who are already interested in fermentation. I found that something is missing; that is educating people and helping them remove the mental barriers to take the first step and give fermentation a green light.

Therefore, this is also a book to help you get into the benefits of vegetable fermentation and to help you conquer the mental barriers that may be preventing you from getting started.

Additionally, this is also a book to teach you how to master vegetable fermentation quickly, even if you know nothing about it at the moment.

In fact, by the time you finish this book, you will have achieved this. If I could master vegetable fermentation starting from knowing basically nothing about it before working on this book (which is a true story), then I am sure you can do that, too.

And, lastly, this book also serves as a gateway to awaken your awareness of the existence and importance of microbes, and to inspire you to continue to explore further how to more effectively co-exist with these important allies, in ways not just limited to consuming fermented vegetables.

Below I am going to share with you how I designed my learning process:

To educate myself on the basic knowledge, I dived into research journals, studies, and books regarding microbiology and fermentation, including vegetable fermentation.

In order to understand what people are concerned about regarding fermented foods and what stops them from trying such foods out; I designed a survey and reached out to more than 100 people to collect their feedback, which helped me better understand what prevented them from trying something that could actually improve their health.

Then, I conducted research based on these concerns and interviewed scientists, fermentation experts, best-selling authors, and health professionals to gather answers for the respondents to try to expel their worries.

Here comes the fun part – how did I teach myself to master fermented vegetables quickly?

I started by reading different fermentation books on the steps required for fermentation; then I deconstructed all the procedures in detail and tried to find consistent patterns from different types teachings. I came to realize that there are only nine basic steps you need to follow.

To master the details of the fermentation techniques, I also got in touch with fermentation experts like Frederick Breidt, professor at North Carolina State University who specializes in vegetable fermentation and who works for the USDA, Sandor Katz, author of the New York Times best-selling *The Art of Fermentation*, Kirsten, the co-author of one of my favorite books *Fermented Vegetables*, from whom I learned a lot on how to make and eat fermented vegetables, and Donna Schwenk, best-selling author of *Cultured Food for Health*.

Afterwards, I pulled together 16 established recipes that are just enough to cover three basic types of fermented vegetables (krauts, kimchi, and pickles), made minor tweaks, experimented on them, documented my observations during the entire process, and jotted down my reflections and learnings along the way, which I am sharing in this book.

Then, this book was born, a book that brings your attention back to the most capable allies on and underneath your skin (that you never want to part with), a book that tells you it is okay to let go of your worries and to give vegetable fermentation a try, and a book that teaches you how to quickly master this type of fermentation even from zero experience.

There are four big sections in this book:

Section I lays down all the fundamentals you need to know before you start. It helps you see the forest before I take you into the trees.

Specifically, you will learn: why you should pay attention to your microbes and know the importance of nourishing the gut (Chapter 1); how you are subconsciously killing your microbial allies and how to help them survive and thrive (Chapter 2); and what role vegetable fermentation plays in this big picture and how consuming fermented vegetables can give you benefits physically, emotionally, and spiritually (Chapter 3).

In **Section II**, you will learn how you can become an expert in vegetable fermentation in six steps.

Step one is to demystify your top concerns, so that you remove your mental barriers which strip away your interest and prevent you from giving it a try (Chapter 4); step two is to list all the most essential items you need to ensure your fermentation experiment is both smooth and safe (Chapter 5); step three is to hold your hand as I go over all fundamental principles of making three basic types of fermented vegetables, with a thorough step-by-step guide that summarizes nine procedures you *always* need to consider for each type of vegetable fermentation project (Chapter 6); step four is to walk you through 12 areas you should pay attention to when your fermented vegetables are done and ready to serve (Chapter 7); step five is a short and sweet chapter that teaches you how to preserve your batches of fermented vegetables (Chapter 8); step six is to guide you to go beyond this book by taking actions and joining communities to let the fun begin (Chapter 9).

Section III is a collection of personal thoughts.

First, I want to invite you to *not* think about vegetable fermentation as the "Holy Grail" and instead share with you my thoughts on how to cope with some potentially negative experiences related to vegetable fermentation (Chapter 10); last but, of course, not least, I write down my final thoughts that go beyond vegetable fermentation, where I will take you away from the trees and invite you to hover above them and see the forest again, as I discuss what I learned from the study of vegetable fermentation and microbiology, and why I think we need to respect ancient wisdom and leverage it with nutritional science and food education—as part of our path to seek out the right way to eat (Chapter 11).

Finally, all my fun experiments are documented in **Section IV**, including how I made my 16 bottles of fermented vegetables and personal lessons learned (Chapter 12) as well as 10 recipes I created to incorporate fermented vegetables into my daily life (Chapter 13).

This book contains a lot of studies and may sometimes overwhelm you. In order not to make you feel stressed and help you digest, reflect, and immediately turn knowledge into actions, I have also included chapter summaries and some easy homework for you at the end of each chapter starting from the Introduction to Chapter 11. When necessary, I have also incorporated checklists which you can download.

Now, let the journey begin.

Chapter Recap

- *The book will help you: a) learn more about your body and microbes; b) understand the role that probiotic supplements and fermented foods play in the big picture; c) conquer your mental barriers regarding fermentation; master vegetable fermentation quickly; and e) use knowledge in this book to continue to explore how to better co-exist with your microbes.*

Exercise

•*My goal of reading this book is to _____.*

SECTION I: THE BIG PICTURE

Chapter 1:
Meet Your One Hundred Trillion Allies

"There are more microbes present on your hand than there are people in the world…Although the microbes that inhabit us were, in the beginning, just looking for food and shelter, over the course of our coevolution they have become a fundamental part of our biology."

– ERICA SONNENBURG & JUSTIN SONNENBURG
The Good Gut

Human beings and microbes have co-evolved throughout human history. For any kind of partnership that exists for so long, there must be a healthy ecosystem in place. Just as Charles Darwin said, "it is not the strongest of the species that survives, nor the most intelligent, but the one most responsive to change."

Through evolution, the microbes who are allies on the skin of and inside the human body must have learned how to co-exist with human cells harmoniously, forming an inter-connected relationship.

If you are owner of a factory that runs a machine called "the human body", without which your daily operation and production could not function properly, then your microbes are a group of your most capable and experienced inspectors and employees to guarantee the machine is rust-free and operates effectively and efficiently.

They are helping you keep this factory functioning and well-maintained, so that you can pay them timely salaries—which is a way they can feed themselves and protect their family members as well. In other words, you need them for optimal operation; and they need you for life-sustaining support.

Microbes, Invisible Forces from Head to Toes

Microbes are single-celled bacteria which represent the oldest form of life on Earth (about three and a half billion years old). In contrast, we humans emerged "only" about 200,000 years ago. In their book *The Good Gut*, Dr. Justin and Eric Sonnenburg help us picture how ancient these little guys are:

If you consider the history of the Earth to date as a 24-hour day, with the planet being created at midnight, microbes would have emerged a little after 4 a.m., whereas humans would not have appeared until a few second before the end of the day. Ancient as they appear to be, they are also very advanced, because through evolution they have learned how to quickly adapt to the exposure to new environments. Therefore, their ability to adjust is likely to be stronger than that of a human being, because they are billions of years older. They find residence in every newborn baby and are able to quickly settle in, becoming dependents for food and shelter as well as becoming a fundamental part of our biology.[1]

In other words, we have grown to such a state of equilibrium that we know how to depend on each other for nourishment, protection, self-defense, and sustenance, just as the owner of the factory and its employees need each other.

You are living in an invisible yet vibrant microbial community, so vibrant that you might feel dwarfed after becoming familiar with the size of this particular community you are living with: do you think 20,000 human genes in the body is a big crowd? How about the two to 20 million microbial genes that are sharing the same habitat?

Also, before you get astonished by the 10 trillion human cells within you, think about the whopping 100 trillion microbial cells that you are carrying.[2] On each human cell, there are 10 resident bacterial cells. No wonder Michael Pollan says: "we are only 10 percent human."[3]

These 100 trillion microbial cells have three pounds of weight, a weight equivalent to that of your brain, and similarly to the brain; it's one of the most important weights in your body that you don't want to shed.

Now that you understand microbes are ancient yet advanced, massive and indispensable forces residing on the skin and in the body, here is one more thing you should know about them: just as you are unique in your own ways, so are the microbes you are living with.

Microbiota is unique to you like your own genes and finger prints. That's because the history of an individual is highly personal: the way this person was born (whether the baby was delivered vaginally, or via Cesarean delivery), medical history (whether this person has been treated with antibiotics frequently, or not at all), specific living conditions (whether a child grew up with pets, plants, or dirt in the garden), the types of foods consumed (whether they are highly processed foods, or naturally grown produce without treatments of artificial hormones or chemicals), and the kind of lifestyle the person is living (whether it is in a fast-paced stressful life at an investment bank, or a mindful life with constant self-reminders to adjust, de-stress, and relax oneself).

These are a few examples that showcase how microbes can sensitively react to human activities and external environments, therefore influencing their presence in human bodies. That is why the microbiota in each and every one of us is unique. We are all responsible for finding a best way to eat and live, that fits our unique bodies and allows us to co-exist with the microbes within us. Only you know what's best for you.

If you understand this, you will quickly learn that consuming fermented vegetables may or may not be your thing, although I can guarantee you that it is one of the safest and easiest ways to expose yourself to some of the most gut-friendly microbes, as you will learn more about later in Chapter 3.

I am not saying this to discourage you; rather, I want to show you the reality that: ultimately, you will have to figure out your own path to maintain, nourish, and enhance the health of your internal microbes, which I will talk about in detail in Chapter 10.

Now that we know all these facts—the history of microbes, their ability to adapt, their presence in great quantity, and their exclusivity and commitment to no one else but you alone, I think this is a perfect time to remind you that it is never only about our own selves, since we are the leaders of some 10 trillion human cells residing in the human body along with 100 trillion bacterial cells, which altogether makes us look and feel small by comparison.

Meanwhile, it is never only about serving for our own good either; instead, we have to serve these microbes to make sure that there is a good amount of beneficial bacteria present, so that we can together make our bodies stronger and healthier, like a room that is clear of dirt and clutter, a castle that is hard to attack, and a machine that always runs with high speed and great efficiency.

To help you go through what you have just learned so far, here is a five-minute-and-28-second animation which vividly describes how the microbes form the "invisible universe" to serve your gut and the entire body. You could visit http://bit.ly/invisible-universe.

If you feel like more education, here is a 17-minute-and-24-second TED talk in 2014 by Rob Knight, pioneer in studying human microbes. Simply visit http://bit.ly/human-microbes.

Gut Flora for Your Digestive System and Beyond

As I did my research, I found out that most of the microbes live in the digestive system. Because of this, and because of the fact that they are an integral part of our biology due to our coevolution, the health of the microbes is crucial for your gut health. Are you currently suffering from digestive issues? If yes, then you may be able to find reasons and solutions from this following section.

To make things more interesting, the importance of gut flora does not just stop at the digestive system, its contribution goes beyond the gut and to the rest of your entire body. In other words, the health of your gut improves digestion as well as your overall well-being; this makes it another reason for which you should not skip this section.

Gut Flora Helping with Digestion

As soon as you swallow down your foods, the digestion process starts to take place. The army of gut bacteria facilitates different tasks after taking the order from the Commanding General, the gut.

Overall, gut bacteria break down the carbohydrate and turn sugar polymers into simple glucose via a type of enzyme called CAZymes (Carbohydrate-active enzymes).[4]

Specifically speaking, here are some examples:

Lactobacillus species in the small intestine produce lactase, the enzyme required to break down lactose (the sugar in milk); the digestion process forms lactic acid, which keeps the "unfriendly" bacteria in check.

Bifidobacteria in the colon produce lactic acid to provide energy required by cells that line the intestine wall and to kill harmful bacteria; meanwhile, they also produce B and K vitamins. Both species create an acidic environment that increases the absorption of minerals such as calcium, copper, magnesium, and iron.[5]

Bacteroidetes digests complex sugars known as xyloglucans, which make up to 25 percent of the dry weight of dietary fruits and vegetables including lettuce, onions, eggplants, and tomatoes.[6]

Before we move on, let's take a moment to acknowledge that your body is a piece of very sophisticated machinery that is built with different systems inside it. As mentioned earlier, your gut bacteria are not just influencing the gut, but also closely interacting with internal systems. And, who knows, perhaps another breakthrough study may be on its way – as you are reading this book – which will reveal even more interesting relationships between gut bacteria and human systems, within unknown or unexplored territories.

Next, we will learn specifically how your gut bacteria are related to brain development and are beneficial for the nervous system, endocrine system, immune system, metabolic system, skeletal system, and cardiovascular system.

Gut Flora for Brain Development and Their Influence on the Nervous and Endocrine Systems

First, let's consider how your gut is affecting your brain to influence your learning, behavior, and memory. Do you still remember an example earlier on how a group of researchers at the University of California, Los Angeles (UCLA) conducted a study and discovered that an increase of probiotics in the gut led to altered brain functions (the periaqueductal grey getting close to the cognition-association areas of the prefrontal cortex which controls decision making, learning, and self-discipline)? This finding points to the same conclusion of another study, too.

A group of researchers from McMaster University in Hamilton, Ontario conducted a study in 2013 and found out that bold mice got timid when they got microbes from anxious mice, and that aggressive mice calmed down when they were fed with probiotics. The researcher also found out that an increase in microbes can cause an increase in a chemical called brain-derived neurotrophic factor, which plays a role in learning and memory.[7]

Further, more and more recent studies have found the direct impact of microbes on both the nervous system and endocrine system. There is even a specific research field dedicated to the study of the interaction of the three, called microbial endocrinology.

From the change in temperaments in mice in the experiment above, you can easily infer that microbes can influence your mental state and your mood, too. To illustrate how microbes exactly change your mood and behavior, here is a very interesting study I found regarding gut bacteria and the neuroendocrine system. Below is an exact citation from the report in 2013 published in the US National Library of Medicine, National Institute of Health:

"This ability of microorganisms contained within the microbiome to influence behavior through a noninfectious and possibly non-immune-mediated route may be due to their ability to produce and recognize neurochemicals that are exactly analogous in structure to those produced by the host nervous system."[8]

As you see, microbes may have the ability to produce neurochemicals almost identical to those your brain produces. Then these microorganism-produced neurochemicals may directly interact with elements of the nervous and endocrine systems, which can lead to changes of your mood and behavior.

You now understand how microbes can influence your mood and behaviors, but in what way? In other words, do microorganisms influence your mood and behaviors in a positive or negative way?

Again, the mice example above might give you some hints already: the fact that aggressive mice calmed down after receiving probiotics suggests the mind soothing properties of microbes.

Looking inside, you will find the gut is full of mind-soothing treasures:

The gut produces about 95 percent of the body's serotonin, the neurotransmitter usually brought up in the context of depression.[9] Further, enkephalins, one class of the body's natural opiates, are in the gut; and on the subject of the gut, there has been another breakthrough finding: it is even a rich source of benzodiazepines – the family of psychoactive chemicals that includes such ever popular drugs as Valium and Xanax, which are used for anxiety disorders and for achieving calming effects.[10]

If that's not enough proof, more and more studies have come to the same conclusion that exposure to microorganisms shows promises for preventing and treating depression in the modern age. They do so by "training the human immune system to tolerate a wide array of non-threatening but potentially proinflammatory stimuli". Lacking such immune training can increase the likelihood of inflammation inside the body, leading to major depressive disorders.[11]

Gut Bacteria to Strengthen Your Immunity

Now that we slightly touch upon the interaction between microbes and your immune system, let's go further and explore together how microorganisms play an important role in strengthening your immunity.

Did you know that about 70 to 80 percent of your immune system's activities are actually happening in your gut?[12] Gut-associated lymphoid tissue (GALT) is the prominent part of mucosal-associated lymphoid tissue (MALT) and represents almost 70 percent of the entire immune system; within GALT, about 80 percent of plasma cells (mainly immunoglobulin A (IgA)-bearing cells) reside right in there.[13]

The gastrointestinal system including endogenous microorganisms plays a very crucial role in maintaining immune system homeostasis.

Changes of these microbes can increase the risk of food allergies or chronic inflammatory intestinal diseases.[14] Microbiota strengthens your immunity by influencing the activity of hundreds of your genes, helping them express their traits in a positive and disease-fighting way.[15]

An interesting breakthrough study in 2010 revealed that changes in the activity of hundreds of genes happened in the human body after individuals consumed a dairy drink containing three strains of probiotic bacteria, which led to changes similar to the effects of certain medicines in the human body, including medicines that positively influence the immune system.[16] Although these effects from probiotics are similar to effects of components applied to medicines by the pharmaceutical industry, the effects from probiotics are less strong and barely have negative side effects.[17]

This is indeed a win-win situation both for the microbes and the immune system as a 2013 study shows that "a cross talk between the mucosal immune system and endogenous microflora favors mutual growth, survival and inflammatory control of the intestinal ecosystem."[18]

The significance of gut microbes influencing your immunity does not just stays within the gut, but even throughout the whole body.

A study published in the journal *Cell Host & Microbe* showed that "beneficial gut bacteria played a key role in the development of innate immune cells – especially macrophages, monocytes and neutrophils – special white blood cells that provide a first line of defense against invading pathogens." These white blood cells circulate in the blood and are stored in the spleen and the bone marrow. It may be due to the circulation of these disease-fighting cells throughout the body that mice with healthy gut flora were able to fight off infection after being injected with the bacterium *Listeria monocytogenes*, which is harmful to humans. At the same time, the germ-free mice failed to fight off the infection and died.[19]

The lead author of this study comments later: "it's interesting to see that these microbes are having an immune effect beyond where they live in the gut. They're affecting places like your blood, spleen, and bone marrow – places where there shouldn't be any bacteria."

So far, you've learned how your gut flora positively affects your learning and memory, benefits your digestive system, influences your mind, mood and behavior, and strengthens your immunity – not only within the gut, but also throughout the body. But, what about your metabolism? Can gut microbes improve your metabolism and help you lose weight? The answer is that it is very likely.

Gut Flora for Weight Loss

Accordingly to Jeffrey Gordon, microbiologist and director of the Center of Genome Science and Systems Biology at the Washington University School of Medicine in St. Louis, microbiota helps control your metabolism.

One study Gordon and his team conducted discovered that the invasion of a group of organisms called *bacteroidedes* were associated with metabolism and the prevention of the increased weight. They also found out it is very likely that this particular group of gut bacteria prefer and rely on a high fiber diet.

From this finding, he concluded that ingredients in the diet, which the bacteria can use, can favor both the microbes themselves and humans, and that what you eat and your gut flora collaborate and together shape properties in the microbial communities that affect the body.[20]

Can microbes alone contribute to weight loss? Absolutely. This was indicated from another study conducted by Gordon and his colleagues.

Two groups of germ-free mice were placed in two separate cages and ate the same diet in equal amounts. The mice that received bacteria from an obese person grew heavier and had more body fat than animals with microbes from a thin person. Then, Gordo and his team moved all animals to a shared cage. Results showed that the mice carrying microbes from the obese human had picked up some of the gut bacteria – especially varieties of *Bacteroidetes* – from the lean counterparts; and both groups remained lean. From this series of experiments, you see that it is possible that healthy gut bacteria can help prevent and fight obesity.[21]

What about gut bacteria's contribution to bone development and prevention of heart disease? These are two areas that you are about to learn next.

Gut Bacteria for the Skeletal System

There is one more alternative to treat osteoporosis now; and I think you can guess what that is already – the use of gut microbiota. Although further studies are still needed, people start to see that the gut bacteria can even become a major regulator of bone mass and can normalize immune status in bone marrow.

During the study, germ-free mice have increased bone mass associated with a reduced number of osteoclasts – a large multinucleate bone cell that absorbs bone tissue during growth and healing, critical in the maintenance, repair, and remodeling of bones of the vertebral skeleton compared with conventionally-raised mice. And colonization of germ-free mice with a normal gut microbiota normalizes bone mass and generates more osteoclasts.[22]

Gut bacteria also improve immune status in bone marrow and can positively affect osteoclast-mediated bone resorption.[23]

All these findings have sparked researchers' interest in continuing to conduct more studies to evaluate the gut microbiota as a new way for bone growth, development, and healing.

But the power of gut microbes does not just stop there. We now will explore their contribution to the health of your cardiovascular system, the last human body system that I want to discuss with you in this chapter.

Gut Flora for the Cardiovascular System

In one short sentence, good gut flora can prevent arterial plaque and heart diseases. Of course, not all gut floras are created equal.

An increased level of the harmful floras that belong to a group called *Colinsella* was found in patients with heart disease and was believed to be associated with lipid accumulation and inflammation in the arterial wall.

On the other hand, the same study also found that gut floras like *Roseburia* and *Eubacterium* were enriched in the group of healthy people and was believed to lead to the possible production of antioxidants by the gut microbiota to help fight inflammation inside the body to prevent heart problems.[24]

The lesson learned is that it is about recognizing the beneficial bacteria and cultivating them inside your gut, which you will learn about more starting from the next chapter.

Another reason for why gut floras can help improve your cardiovascular system is that they are a rich source of vitamin K, which is crucial to prevent heart diseases. Vitamin K contributes greatly to the formation of matrix Gla-protein (MGP), which serves as an important inhibitor of calcification of arteries and is synthesized in a vitamin K-dependent environment in smooth muscle cells. In other words, optimal vitamin K levels are needed to produce a proper amount of MGP to prevent arterial calcification, or hardening of the arteries.[25]

Many of the probiotics-rich foods, such as fermented vegetables, are an excellent source of rich vitamin K. To go down a little deeper, vitamin K2 out of the vitamin K family is synthesized by intestinal bacteria; this means gut floras play a key role in generating vitamin K2 for the body. In fact, it is believed that fermented foods have the highest concentration of vitamin K found in human diets.

For example, natto is a type of fermented soy product served as a staple food in Japan for over 1,000 years. Circulating vitamin K2 concentrations after you consume natto would be about 10 times higher than those of vitamin K1 derived from spinach.

Putting Everything Together

As you see from the myriad of research and studies, what your gut can do is way more than just digestion and assimilating nutrients, it is also closely connected with your brain, intellectual thinking, learning, behaviors, mental states, mood, immunity, metabolism, bone health, and your heart. Who knows, since the trend of studying microbiota is rising, more findings may point to how they can benefit you in other different ways.

There is one thing we can know for sure: the scientific discoveries show that cultivating gut floras is the right direction to promote health and prevent diseases.

The unsung heroes – all the beneficial gut flora – should be brought to our awareness. They are the soldiers who fight relentlessly for our bodies; and they are the one crucial factor rarely talked about these days that greatly contributes to your overall well-being.

You Have Two Brains

In the last part, you have learned a little about how the gut promotes brain development. As a matter of fact, the significance of the gut goes much further than that. It is now considered as important as the brain. That's what I want to talk more in detail here.

Have you ever wondered why you get "butterflies" in the stomach when you get nervous before you are about to make a speech on stage? One reader told me that whenever he felt stressed and anxious, he would experience a worse case of constipation. He said, "I have found that a calm, unhurried, and smiling frame of mind benefits the large intestine in my case." Besides, did you know that anxiety may also cause nausea, pain, tightness, and a general feeling of unease in the stomach?

If you ever wondered exactly why your mental state seems to connect with your gut, you will be able to get answers soon. It concerns how the brain and gut are very closely interacting with each other.

Before I overwhelm you with the wealth of details that follow, let me share with you this simple short conclusion I have come up with: your gut is a "**second brain**", **complex** and tightly **interconnected** with your primary brain; yet it is an **independent** system. "Second brain", "complex", "interconnected", and "independent" are the keywords. Let's dive into these four concepts one by one.

Gut as a "Second Brain"

In recent years, there has been mounting research that shows the enteric nervous system mirrors the central nervous system.

Dr. Michael Gershon is a professor of anatomy and cell biology at Columbia-Presbyterian Medical Center in New York who is considered one of the founders of a new field called neurogastroenterology, a study that focuses on the brain, the gut, and their interactions. According to Gershon, the gut is described as a second brain because the gut shares very similar substances to what's in the brain, including neurons, major neurotransmitters like serotonin, dopamine, glutamate, norepinephrine, and nitric oxide, brain proteins called neuropeptides, and major cells from the immune system.[26]

Since a lot of major substances are in both the brain and the gut, it is not surprising to see the gut can perform a lot of the tasks that you originally thought only the brain could do.

For example, both the brain and the gut regulate your body's immunity at the same time. Studies show that there are many hormones, neurotransmitters, and neuropeptides released by the brain or the structures controlled by the brain to regulate the immune system.[27]

Meanwhile, you will never overlook the gut's ability to keep you away from diseases after you know that the gut's immune system has 70-80 percent of the body's immune cells, suggesting that a lot of disease-fighting activities happen in the gut and that the gut can play a very crucial role in keeping the body strong and healthy.

Gut as a Complex System

Because the gut and the brain share so many similarities, it is easy to infer that the gut must have a complex system to run like the brain, as well.

Your gut's brain consists of tissues lining the esophagus, stomach, small intestine, and colon. On one hand, it is a single entity with a network of neurons, neurotransmitters, and proteins that enable you to act, respond, learn, and remember; on the other hand, it has sensors for sugar, protein, acidity, and other chemical components to facilitate the digestion progress, and determines how the gut should mix and process the foods you eat. Just as Gershon says, "[the gut is] not a simple pathway; [and] it uses complex integrated circuits not unlike those found in the brain."[28]

Gut-Brain Connectedness

How are your gut and brain connected?

Did you know that the brain and the gut are created from the same type of tissue called the *neural crest* during fetal development? Then, as the fetus grows, one part of the tissue develops into the central nervous system, another part into enteric nervous system to govern gastrointestinal functions with its network of neurons; and they are connected via the vagus nerve – the tenth cranial nerve from the brain stem down to the abdomen.[29]

The connected relationship is reflected in two ways: the brain influences the gut; and the gut influences the brain, too.

Since the brain and gut are connected via the vagus nerve, the brain helps regulate gut activities by sending off signals via nerve fibers down to the gut. For example, the brain initiates responses to feeding even before the ingestion of food.

Imagine you are flipping through a food magazine and see the juicy food photographs with vibrant colors that almost pop above the pages as if these dishes were delivered right to your mouth; or, when you walk by a neighbor's door and smell a faint aroma of sweet and sour barbecue sauce. During either situation, you may start experiencing the feeling of hunger, a grumbling noise inside the stomach, or secretion of saliva. This is because the very sight and smell of food stimulates exocrine and endocrine secretions in the gut and increases gut motility.[30]

Now you know a little bit about how your brain influences the gut. So, what about the other way around? How does the gut influence the brain?

To explore this, a group of scientists at the UCLA, conducted a study and discovered that women who regularly consumed beneficial bacteria known as probiotics through yogurt demonstrated "altered brain function", both in a restful state and in response to an emotion trigger.

During the study, women who consumed probiotic yogurt showed greater connectivity between the key brainstem region known as the periaqueductal grey and cognition-association areas of the prefrontal cortex, the part of the brain that is crucial for learning, self-control, planning complex cognitive behaviors, decision making and taking actions. The women who ate no dairy product at all showed greater connectivity of the periaqueductal grey to emotion and sensation-related regions.[31]

By proving the gut directly influences the brain, the study shows that the consumption of probiotics can help you become calmer, develop more self-control, and improve your learning skills, which amazes a lot of researchers and me as well.

Gut as an Independent System

Not only is the gut is closely connected with the brain, it also has its own universe independent of the brain, too. There are more than 100 million neurons; yet, the vagus nerve sends off signals via only a few thousand nerve fibers to the gut. Interestingly, about 90 percent of the fibers in the vagus send messages from the gut to the brain and not the other way around.[32] All this suggests that the gut has its own separate universe, not directly controlled by the brain.

For example, it has its own system to facilitate the entire digestion process including breaking down food, absorbing nutrients, and expelling waste which involves chemical processing, mechanical mixing and rhythmic muscle contractions. The gut performs this individually. That's why you don't have to remember to remind the gut to digest and absorb nutrients when it's the time; nor do you need to control how food should be passed through the intestines. Your gut does that for you already.

A Reflection on How the Study of Microbiology and Traditional Chinese Medicine Converge

As I was doing research, my inner-nudge urged me to create this extra bit of personal insight: the modern study of microorganisms has silently connected us back to traditional wisdom.

Since 2014, I've been studying Traditional Chinese Medicine (TCM) and Chinese Food Therapy; and I've found out that many of the scientific studies on microorganisms and much of what TCM and Chinese Food Therapy have advocated for over 2,000 years points to the same directions and conclusions, which I will point out with the following examples:

Example One: The Health of Your Stomach is Crucial.

According to TCM, five elements – metal, wood, water, fire, and earth – are the foundation for everything in life; and each organ inside the body belongs to one element while carrying its characteristics. And, TCM sees that the stomach belongs to the element of "earth", the most important element in life.

Based on *Inner Canon of Huangdi*, an ancient Chinese medical text that has been treated as the foundation for Chinese Medicine for over 2,000 years, earth (or, soil) nourishes everything in life, and therefore, is the source of life. By connecting the stomach to the element of "earth", this classic medical text suggests that nourishing the stomach is vital to sustaining and prolonging life.

On the science side, numerous scientific studies including the study of microorganisms have elevated the importance of the role of the stomach, too.

One of the major functions performed is defense. The stomach kills most foreign organisms that you ingest with highly acidic stomach fluid secreted inside the organ, so that they cannot get into your bloodstream or anywhere else in your body.[33] Besides, as covered earlier in this chapter, the stomach is also a place which hosts microbes that can benefit your brain and the majority of the life-sustaining systems inside you.

From my research, I even found a story from a Chinese microbiologist named Liping Zhao, who also realized the connection between the TCM five-element theories (e.g. stomach belongs to "earth" element and is central to health) and the study of microorganism. Because of this, he decided to adopt a hybridized approach to improve health by consuming more whole grains – which TCM believes can nourish the stomach, and which are rich in prebiotics fibers important for the beneficial bacteria – along with other healing traditional medicinal foods like bitter melon and Chinese yam.

The result? He lost 44 pounds over the course of two years.[34]

Example Two: Your Spleen Also Plays a Vital Role in life.

As I mentioned before, in TCM each organ inside your body belongs to one of the five elements while carrying this particular element's characteristics. So then what is the element that the spleen is connected to? The answer is your spleen also belongs to the element of "earth", another crucial organ that serves as the foundation for a healthy and vibrant life.

TCM believes that the spleen facilitates circulation of nutrients to the rest of the body; this organ also helps circulate the water inside the body and the blood. A weakened spleen can cause fatigue, dizziness, diarrhea, weight gain, poor digestion, bloating, dull skin, or a pale complexion.

Once thought to be an unnecessary bit of tissue, the spleen is now stepping into the limelight, as mounting research has validated and justified its significant contributions.

The study of microbiota reveals that microbes circulate within and linger in the blood, the spleen, and bone marrow, which is essential to help your immune system fight off infection. As a matter of fact, it is now regarded as an organ where important information from the nerves reaches the immune system. Below is a powerful quote from Science Daily:

"A hundred years ago, the spleen (located in the upper quadrant of the abdomen) was thought to be only a reservoir for blood. It has only been in recent years that scientists discovered that the spleen is a manufacturing plant for immune cells, and a site where immune cells and nerves interact. The spleen defends the body against infection, particularly encapsulated bacteria that circulate through the blood."[35]

Example Three: Different Parts of the Body Are Closely Related.

TCM recognizes that the body is a tightly connected unity and each body part is connected with one another. For example, gastrointestinal problems like constipation can lead to acne breakouts. If you relieve yourself properly, the chances are that your skin will be improved.

In addition, better understanding of microbes has led to further knowledge of how one part of the body can connect to another. Take the example of the increase of awareness that gut microbes can heal acne. *Lactobacillus acidophilus* cultures were mentioned by other physicians in the 1930s as a popular internal means to treat acne. By cultivating a mixture of this particular bacteria and *Lactobacillus bulgaricus* in acne patients, 80 percent of them could see improvements, particularly in cases of inflammatory acne.[36]

Example Four: Body and Mind Are Connected.

Influenced by Taoism, TCM has its spiritual side, too: TCM recognizes that mind and body are closely connected.

TCM believes that the mind-body balance is the key to optimal health and well-being; it also prevents diseases and slows down the aging process.

My Chinese doctor, Fang-Tsuey Lin,[37] has been in practice since 2001 and has her own clinic based in Lexington, Massachusetts. She taught me that if I heal my body, my emotional states and personality will change; if I get angry, anxious, or stressed out, my body will be negatively affected.

How does science come into play to reach the same conclusions?

Earlier, you learned about a group of the UCLA researchers who discovered that the consumption of probiotics-rich yogurt could change brain functions in humans. You have also learned how bold mice become timid after receiving microbes from anxious mice; and aggressive mice calmed down when fed with probiotics.

All these point to a conclusion that: by feeding beneficial bacteria to the body, there is a chance that you will become less emotional, reactive, and more calm; and act with more awareness and self-control.

In other words, these examples showcase how physical changes in the body are connected with your mental states.

Putting it All Together

You have just tasted a little flavor of how traditional wisdom and modern science intersect: they both point to truths about how to understand and take care of the body, so that you achieve optimal health.

This is not to try to prove that TCM is right, after all TCM evolved from its own unique era and is using a system different from frameworks used by science. Rather, I bring this up to call to your awareness the idea that it's OK to look back to traditional wisdom as a reference. As Michael Pollan says in his book *Food Rules*:

"I believe that there are other sources of wisdom (other than food science) in the world and other vocabularies in which to talk intelligently about foods. Human beings ate well and kept themselves healthy for millennia before nutritional science came along to tell us how to do it; it is entirely possible to eat healthily without knowing what an antioxidant is."[38]

It is wise to give weight to what the elderly or older generations have to say about food wisdom and taking care of the body. As you will learn in Chapter 3, although the art of fermentation is the rising star these days, it is actually a very traditional practice that originated from tribes around the world (from China, Russia, Japan, India, Rome, Bulgaria, Ukraine, and more). Modern science has connected us closer to tradition. And I think we should leverage both to best take care of our own health and well-being.

Before we move on, I'd like to end this chapter with an inspiring quote from Sally Fallon, author of *Nourishing Traditions* and president of Weston A. Price Foundation (a non-profit organization dedicated to restoring nutrient-dense foods to the human diet through education, research and activism):[39]

"Technology can be a kind father but only in partnership with his mothering, feminine partner—the nourishing traditions of our ancestors... The wise and loving marriage of modern invention with the sustaining, nurturing food folkways of our ancestors is the partnership that will transform the Twenty-First Century into the Golden Age."[40]

CHAPTER RECAP

- We are co-existing with 100 trillion microbial cells that perform life sustaining functions. Therefore, we need to acknowledge and respect their contributions. They are ancient yet advanced; each and every one of us has our own unique microbiota.

- The health of your gut has a positive impact on many parts inside the body: digestion, brain development, neuroendocrine system, immunity, metabolism, bone health, and heart functions.

- Your gut is a **"second brain"**, **complex** and tightly **interconnected** with your brain; yet it is an **independent** system.

- Both Traditional Chinese Medicine and nutritional science have both simultaneously pointed to many strikingly similar conclusions on how we should take care of the body. This suggests that we need to take advantage of traditional wisdom and food science at the same time to optimize our health and well-being.

Exercises

- There are ____ trillion human cells and ____ trillion bacterial cells in my body. Most of the bacterial cells live in my _____ system.

- List three ways of how healthy and happy microbes are doing good to my body: 1) _____; 2) _____; 3) _____.

- Protecting my _____ is particularly important because that's where most of my microbes live.

- Think about one health issue you are currently most concerned with; then, ask yourself a very simple yes or no question: can my *being nicer to my microbes heal my body and help resolve my health concern?*

Chapter 2: The Civil War and Peace Agreements

"Excessive use of antibiotics, in particular in children, has been shown to be associated with, again, risk factors for obesity, for autoimmune diseases, for a variety of problems that are probably due to disruption of the microbial community."

– JONATHAN EISEN

Now that the existence and importance of friendly microbes has been brought to your awareness, let's take a closer look at how we are currently living our lives and can live better by co-existing with them in a more microbe-friendly way.

As you see, the awareness of these massive invisible forces leads to the realization that there is in fact a big invisible battle happening within us due to the modern age.

It is like you are the CEO of the factory keeping a team of the most capable employees who hope to serve and help maintain optimal operational efficiency, yet we refuse to give them the food they need. We deprive their sleep, and strip away their working energy when we, ironically, want to run the factory properly and produce quality products in return.

The Civil War – Ways We Are Threatening Our Allies

There is a fierce battle going on right inside us. The war is perpetuated by the way babies are born and brought up these days, industrialized processed foods, widespread use of antibiotics, modern lifestyle, and overuse of antibacterial chemicals in your daily life (or, your obsession with cleanliness) all contribute to the killing of microbes. I will now cover them one by one for you here.

The Problem of Caesarean Deliveries

When I was in my early twenties, I remember one time I was having dinner with my girlfriends at a local restaurant. There were about 10 of us sitting at a round table, a typical one with a round spinning tray on top (commonly seen at Chinese restaurants) to serve large groups of customers. The topic of becoming a mom and giving birth to a child came up randomly during our talks.

One girl suddenly raised a question: "if you were to give birth to your child, would you choose naturally giving birth or opt for a C-section?" I remember that scene so vividly that 50% of the girls voted the latter due to personal reasons, like wanting to minimize pain. I chose naturally giving birth, because I thought that it was nature's gift to women to deliver a baby through the birth canal.

The poll back then was only based on a hypothetical circumstance and didn't reflect what we ended up deciding to do. Interestingly, the ratio of 50% stuck in my mind; and I later found out that the current Chinese caesarean section rate is just about 50%, way above the recommended level of 15% by the World Health Organization.[1]

Based on the study, many Chinese women choose to do C-section due to personal reasons, rather than health ones. Besides that fact, with C-section rates from many other countries also going beyond 15%, we see that there is a trending demand for C-sections. Based on what I was told when I was in China, C-sections can reduce the chance of the body losing shape, which is why celebrities in Asian countries choose it; and it takes less time while causing less pain.

But is it really an ideal decision?

I later found out that performing a C-section is one way to threaten the lives of the friendly microbes.

Vaginally delivered babies have more *Lactobacillus* and *Bifidobacteria*, two crucial groups of gut flora that are key to digestion and metabolism, as discussed in the last chapter. In fact, one recent study found out that cesarean delivered babies were colonized by a mixture of potentially pathogenic bacteria typically found on the skin and in hospitals.[2] Besides, there is an increasing body of evidence that the intestinal microbiota play an essential role in the postnatal development of the immune system.[3]

All of the above is in line with the discovery that shows a C section baby is more vulnerable to health issues like asthma, food allergies, hay fever, and obesity later in life.[4]

Now, do not go extreme here. I am not saying pregnant ladies should *all* opt for vaginally giving birth to babies at all cost. While acknowledging the benefits, the pregnant moms should not choose vaginally giving births blindly; and C-section has its own benefits, too, to "deliver a healthy baby and to protect the health of the mother", as the book *The Good Gut* mentions.[5] But one should really balance the pros and cons before making a decision.

The Missing Link of Breast Feeding

Another reason that human microbes are being threatened is the lack of breast feeding to babies these days. Human milk provides complete nutrients to the baby including fats, protein, carbohydrates, and other health promoting compounds; and it strengthens the baby's immunity as well. It also contains something called human milk oligosaccharides, or HMO, which is a collection of complex carbohydrates and the third most abundant class of molecules in breast milk after fat and lactose.

Interestingly, HMO is nature's creation made not for the baby, but for feeding the microbes like *Bifidobacteria*, which is crucial for a baby's gut health, and for seeding *Bacteroides*, an important class of microbes that helps the baby to transition to solid foods later on.

Why can't regular milk formula deliver the same benefits as HMO? Well, as a matter of fact, milk formula companies who are aware of the benefits of microbiota in babies have started to manufacture carbohydrates to mimic the structure of HMOs or add living bacteria.

However, these are just not the same. As the book *The Good Gut* points out, "HMOs are human specific; no other animal makes a mix of carbohydrates exactly like them." Scientific research and nutritional engineering about baby formulas is only less than a hundred years old, whereas breast milk comes from thousands of years of human evolution.

In other words, nature knows what is best for babies. Scientists are still decoding human milk components and exploring what specific probiotic bacteria are ideal for babies; therefore, it is not guaranteed that the added probiotics into formulas are really what the baby needs, although these added probiotic bacteria are from a good intention to improve a baby's microbiota.[6]

What's Wrong with Industrialized Processed Foods

Personally, I consider processed foods as a nutrition-depleted ghost town that sidetracks you by pure sensations: they look enticing because of eye catching packaging and artificial coloring; they taste irresistible due to artificial flavorings and texturants (chemicals that enhance textures to foods); and they maintain the same flavors and looks due to addition of unnatural preservatives, all of which are additives that can lead to a series of health issues like cancer, allergies, and brain damage.

Despite the fancy opening presented to you by this ghost town, you will barely find natural fibers, enzymes, vitamins, and other nutrients; and this lack of nutrition can lead to a weakened digestive tract and harm your internal beneficial gut flora.[7] As you will learn later in this chapter, dietary fibers are particularly important for feeding gut bacteria. Consuming too many processed foods, which are typically low in fiber,[8] is just another way to starve your friendly allies that want to make you healthy and strong.

Typically, processed foods are high in high fructose corn syrup (HFCS). Next time when you pay attention to reading labels as you go grocery shopping, you will easily find that HFCS is often labeled in different processed foods.

This additive can damage intestinal lining, and cause toxic bacteria and undigested protein to gain access to the bloodstream, leading to infection and inflammation.[9] According to *The Good Gut*, the gut lining is protected by a layer of mucus which serves as a rich source of carbohydrates that some bacteria within microbiota can eat.[10] Therefore, a leaky gut inevitably threatens some microbes by taking away their mucus food source.

Besides, since the gut is where the majority of the microbes reside, hurting the gut can hurt microbes, too.

To add insult to injury, processed foods are also loaded with trans-fats and processed vegetable oils. Excessive intake of trans-fats is linked with alternation in the gut microbiota.[11] Modern ways to produce fats in the form of hydrogenated oil and trans-fats (often disguised as "partially hydrogenated oils") can cause oil to be prone to going rancid and oxidizing, which promotes inflammation. According to Nourishing Traditions, consuming this type of processed oil can lead to health concerns like cancer, obesity, immune system dysfunction, and bone and tendon issues.[12]

Do You Take Antibiotics Too Often?

"Anti" is the Greek word for "against"; and "bio" is for "life". Putting these two words together you will get antibiotics as the agent to kill living beings. They are intended to use when a human body is infected by harmful bacteria; and we should all be thankful for the fact that the use of them has saved countless lives.

On the flip side, as the word suggests, they kill all life forms, not just limited to pathogens. When antibiotics are used, good microbes are wiped out as well. The current perception that bacteria are all germs causing trouble to the body makes people believe that taking antibiotics is the answer to keeping the body bacteria-free and healthy.

By now you have understood that there are trillions of protective microbes that are crucial for your health and well-being; and the overuse of antibiotics might cause more harm than good over time. The use of antibiotics directly leads to profound and immediate decimation of the gut microbial community, which means that your immune system and other body systems will be compromised.

Another important thing to note is that microbiota may recover over time if you treat the body properly later on; but the microbial profile will likely never be the same.[13] This reminds me of a dialogue between me and one Traditional Chinese Medicine (TCM) practitioner

I visited in 2014, who has inspired me so much to look into TCM as a holistic wellness practice. During our first consultation session, she asked whether I used to take any antibiotics. I told her: "I have taken antibiotics once," immediately followed by an emphasis, "but only once!", thinking that it may help alleviate the negative impact on my body pictured in her mind.

She gentled shook her head, smiled, and replied: "even if you've only had antibiotics once, it is enough do harm to the body that lasts for a long time."

Having said that, we should still stay positive about the pros that antibiotics have brought into our life; but at the same time, we also need to evaluate the down-sides antimicrobial drugs can bring to your friendly microbial community and your body, should you not be careful about taking them.

Too Stressed Out? Lacking Sleep?

The modern lifestyle has brought forth a chronic stress we experience every day. This is a classic example that demonstrates how the state of your mind can influence your overall physical health.

For example, the stress that you are experiencing daily is linked to a decrease in nutrient absorption, decreased oxygenation to your gut, decreased enzymatic output in the gut, and decreased metabolism because of four times less blood flowing to your digestive system.[14] The act of chronic stress hurting your friendly microbes inevitably increases your chances of suffering all kinds of diseases, such as a variety of gastrointestinal problems like small intestinal bacterial overgrowth (SIBO), inflammatory bowel disease (IBD), irritable bowel syndrome (IBS), and a leaky gut which leads to inflammation in the body.[15]

Stress also influences your sleeping quality. This can bring you into a vicious cycle: the poorer your sleeping quality is, the more stress you have because you are anxious about finding solutions for your insomnia, which may lead to another sleepless night. The body needs you to be in deep sleep mode to be able to heal itself; if you are sleeping less than seven or eight hours a day, you might be preventing your body from repairing internal gut tissues.

Ancient Chinese believe that putting the body in the deep sleep mode between 10pm and 5am the next day is crucial to activate the body's self-healing mechanism and to better prepare for the next day. That's why some people also consider the sleep in these seven hours as your "beauty sleep".

Yet, these days, people are likely still enjoying TV, browsing the web, playing with mobile devices, and responding to emails at 10 o'clock at night, unless you have kids to take care of and you need to wake up early to adjust to their schedules and daytime activities.

Technology has definitely brought us more excitement and stimulus to stay up late at night; however, it has also lowered the volume of our microbial channel and weakened our ability to sense signals sent by the body to wind down and take a rest. In the days when we are busy looking outside, we forget the importance of looking inside.

Are You Too Obsessed about Being Clean?

A recent article published by BBC points out that being too clean is wrong. These invisible negative forces come from chemical-filled detergents and cleaners, the chlorinated water, and antibacterial soap which the gut is very sensitive to. Excessive cleanliness has caused the antimicrobial resistant bacteria (or, "superbugs") to thrive: increasing in numbers more quickly than antimicrobial susceptible bacteria.[16]

Although it is not determined how much use is too much, it is a good start to know that you don't have to be obsessed about excessively cleaning your skin, body, and living environment.

For example, just because you pick up a rock and throw it into the trash can, it doesn't mean you have to thoroughly wash your hands with antibacterial soap all the time; just because you go out shopping the whole day, it doesn't mean that you have to use hand sanitizer every one hour; just because you need to wash dishes to keep them clean, it doesn't mean that you have to wash them with antibacterial dishwashing detergent for every single one of the containers.

It's advised to use antibacterial agents with awareness and your own judgment, to moderate the usage, and to use them only when you think it is necessary while keeping in mind the well-being of the friendly microbial allies.

As these invisible battles – the rise in C-section cases, the lack of breast feeding, consuming industrialized processed foods, overuse of antibiotics, chronic stress, lack of sleep, and excessive cleanliness – are actually quite intense right now. You may find that most of the issues addressed here are somewhat related to you, your spouse, or your other family members. But the good news is that the number of risks you can be exposed to matches the number of ways that you can improve the quality of life for you and others.

When I was first awakened by all these alarming messages, I was overwhelmed, too. As I am writing this, there are still things I am not perfect at doing and taking good control of. But I am definitely becoming more aware of the exposures that can harm the good microbes inside me and my family members, while slowly taking steps to minimize the prevalence of gut-threatening environments. This process takes time, patience, mindfulness, and, of course, discipline (because those bags of chips on the rack can be too enticing to resist).

So don't fret, but remain hopeful. As the Chinese saying goes, behind every challenge lies an opportunity. The eye-opening studies regarding the importance of microbes and how we are treating them gives us a wake-up call. At this point, the yearning for living a holistic and healthy lifestyle is on the rise globally. Isn't this – respecting and learning how to co-live with microbes – a great way to start reorganizing your overall lifestyle? Considering how you eat, sleep, live, and deal with stress are all intertwined with microbial activities within you.

Below are seven ways that I would like to discuss specifically how you can take actions to help your microbial community thrive.

1. It All Starts from Babies

The following section directly speaks to all the pregnant ladies reading this book. Even if you find this may not apply to your situation at this point, knowing this can help you prepare yourself in the future, your spouse, other family members, and other people around you. So don't skip this.

As I am writing this, I am not yet a mom; but I want to be able to experience the feelings of all flavors a mother need to go through one day, such as the pain of giving birth to a baby, the joy of holding a little peanut in my arms, my heart melting when I see its first laughter, and a sense of pride when I see it make its first stand on the bed. Knowledge here is the accumulation of my best knowledge collected so far. And, I feel that I am not only writing this for you but also talking to my future self at the same time.

Since the first batch of microbes a new-born baby is exposed to is crucial and the interactions between gut bacteria and infants can have a huge impact for their life as they grow up, it is important to ensure that the growing process of a baby is accompanied by friendly microbes as soon as the infant greets the world. Specifically, there are three things that a mom can do: vaginal delivery, breast feeding, and weaning with microbes.

Yes, vaginal delivery is important because the microbes a baby is exposed to is particularly crucial for its overall well-being later on, especially its gut health. But what if vaginally giving birth is not an ideal option recommended by the doctor for the sake of the health of the mom and the baby?

This case happened to Rob Knight, a professor at the University of San Diego and expert of determining where different bacteria live. He had a child born by C section. Understanding the importance of microbes himself, the parents decided to use vaginal swabs from the mom and inoculated their daughter at multiple body sites "to ensure that she was exposed to the bacteria she would have encountered had she gone through the birth canal", as the book *The Good Gut* describes.

While this might sound a bit extreme to you – and, indeed, it is still not a mainstream practice – the Sonnenburgs, authors of this book, believed that it is possible this practice may happen more often in the near future, given the rise of awareness in the importance of microbes and of vaginal delivery itself.[17]

Moving onto breast feeding. The number one rule is that a mom can breast feed a baby with however much milk she has. It is recommended that ideally she can breast feed her baby exclusively for the first 6 months; then continue to breast feed in combination with a diet of gut-bacteria friendly solid foods (such as those rich in dietary fiber), for another 6 months. Nightly nursing is also a great way to help the baby develop healthy microbes as it grows.

Now that we have talked about how science has proven developing healthy gut flora is important for babies as they grow, here is an anecdotal story from Donna Schwenk, author of *Cultured Food for Life*.

Schwenk gave birth to one of her daughters Holli who was born seven weeks premature and had stopped nursing when she was 10 and a half months old. As soon as she stopped nursing, she immediately experienced frequent colds and countless sleepless nights. When Schwenk started to feed one to two teaspoons of kefir, a type of fermented milk rich in *Bifidobacteria* and *Lactobacillus*, two species of bacteria mentioned earlier that are crucial for a baby's healthy development, Holli's symptoms were relieved dramatically and she quickly became healthier day by day.[18]

This suggests that if there are times when breast feeding in not possible, feeding the baby with fermented foods and beverages abundant with gut-friendly microbes seems to be a great alternative.

What about weaning? Weaning process should focus on introduce gut bacteria friendly foods rich in dietary fiber. Do not forget to educate to your kids the importance of such a practice as they grow, while also walking the talk by feeding yourself with good microbes and consuming dietary fibers. Here is a checklist of foods rich in dietary fiber I have created which you can check out by visiting this website: www.tracyhuang.me/fvm-resources.

2. Cultivating a Microbial Living Environment

There are two big groups of microbial communities you should cultivate. I personally call one group "permanent residents (PRs)"; and the other, "transient communities (TCs)". PRs are the 100 trillion cells co-existing with you, whereas TCs are the probiotic bacteria you feed the body – such as the probiotic supplements and fermented vegetables that you introduce externally who visit, stay for a while, and then exit.

With a massive troop stationed in the body and responsible for a myriad of life-sustaining functions, it makes perfect sense to take good care of the PRs. You can achieve that by eating dietary fiber on a daily basis. Dietary fiber is indigestible by the small intestine and will eventually travel down to the large intestine to feed the microbes residing in there. Through microbial fermentation, short-chain fatty acids are the products and are believed to contribute to weight loss, lowering inflammation, and helping stop a variety of western diseases.

What about the TCs? If they come and go without staying permanently anyways, why bother feeding them? It is like some people questioning the need to make the bed. If we use bed sheets, blankets and pillows every night anyways, why bother making the bed every day to keep everything in order and make it look tidy? Well, the result of keeping the bedroom tidy, even though it lasts only about 15 hours before you use the bed again, can give you mental peace and clarity and may even help you de-stress. As you see, a temporary thing doesn't mean it doesn't provide value. It is the same case with feeding the body with TCs.

It turns out that, during the short period of stay in the body, these visitors have already done a lot of work for your body, primarily increasing its immunity. First, the body might mistreat them as pathogens and therefore activate defense systems that would normally be turned off against harmful bacteria.

Further, inside our intestinal cells there is a special protein to seal the space between cells to create a barrier to keep pathogens or undigested foods from unexpectedly traveling into the body. The introduction of temporary microbial visitors can stimulate the intestinal cells to produce more protein, creating a sturdier gut barrier and forming a stronger defense.

The process also stimulates production of mucus, a slimy layer on the top of the intestinal wall, for protection from invaders. Not only that, a special group of molecules called defensins is also released and serves as protective agents fighting against invading bacteria, viruses, and fungi.

What are TC candidates? Consider fermented foods like fermented vegetables (which is what I will talk about with more depth later in this book), fermented beverages, miso, tempeh, and sourdough. You can also try replenishing microbes by consuming probiotic supplements.

A question that is often brought up is whether consuming fermented foods or probiotics supplements is better. A short answer to this is that there is no absolute answer, as long as you are sure that the probiotics supplements are produced by a reputable manufacturer and that fermented foods are properly made. You can find more details in Chapter 4 about why I actually prefer fermented foods over probiotics supplements to improve overall health.

Understanding the relationship between PRs and TCs and the importance of cultivating both is important to help you approach the later content in this book. This understanding helps you reflect and see how consuming fermented vegetables is tied into the helping-microbial-communities-thrive big picture: consuming fermented vegetables is one way to keep your microbial community health and robust.

3. Getting into the Habit of Reading Food Labels

What you eat can largely decide the kinds of microbes you have. Therefore, besides paying attention to what you should eat, you need to also watch for what to avoid.

The next time when you shop at a local grocery store, be sure to stay aware from the microbe-killing ingredients, such as high fructose corn syrup, trans fats (usually disguised under the name of "partially hydrogenated oil"), or names which sound chemical to you with vocabularies that are hard to pronounce, such as butylated hydroxyanisole (BHA) and butylated bydroxytoluene (BHT), both of which are often found as preservatives in food industry and ranked "moderate to high" in the "Overall Hazard" level by the Environmental Working Group.[19]

Now here is something you should always do – cultivate a mindset of a curious student who is eager to learn by checking out organizations with consumer health in mind, and using their information resources as your research reference. The non-profit mentioned above, Environmental Working Group, is a good go-to place that I visit frequently for more education. You can visit their site at www.ewg.org.

4. Watching Out For the Use of Antibiotics

In addition to limiting the access of processed packaged foods with harmful ingredients into your mouth, also be aware of the profound antibiotic side effects and try to minimize intake if possible. As an example, when you are experiencing skin issues like acne and are prescribed with antibiotics by the doctor, think twice before you buy them. There must be a way to heal your skin by healing the body internally from a deeper level for your long-term good.

Of course, there are times when consuming antibiotics might be necessary, when the body is attacked by an army of pathogens. We might not always be able to get what we need or want, even though we have the best knowledge and intention; luckily we can constantly and consciously improve overall conditions to get the best results possible.

When taking antibiotics becomes necessary, make sure that you find ways to replenish good microbes at the same time. For example, always make sure that you are feeding gut-friendly fiber, if situations allow, and constantly cultivating good microbes by either consuming fermented foods or probiotics with an established reputation.

Here is the catch: some antibiotics can react with some fermented foods like alcohol. That's why I also recommend you work with a health professional to see if you can eat a certain fermented food while taking antibiotics.

Not knowing the importance of vaginal delivery at the time, the Sonnenburgs took another approach for their C-section baby who subsequently received antibiotics. For two weeks after they took her home, they sprinkled the contents of commercially available *Lactobacillus GG* capsules into her mouth; and, anecdotally, she did not experience some of the health issues that infants in her situation would sometime encounter.[20]

Although microbes attacked by antibiotics may not be able to return to their original state, it is still very hopeful that you can cultivate a much improved gut over time, if you are mindful about feeding and replenishing beneficial microbes to the body, based on scientific studies and anecdotal stories I have encountered.

5. Reorganizing Your Current Lifestyle

Here are two things you can do to improve your quality of life with more energy and peace: getting more sleep and meditation.

With the invention of electronic devices such as televisions, computers, and mobile devices, it is easy to get sucked into a favorite TV show, social media app, or interesting websites before you are aware that many hours have gone by. Besides, swiping the screen of mobile devices while in bed right before you sleep may most likely have become an evening ritual.

All these activities only continue to activate the brain and make the body tense; in fact, even responding to one single email can induce stress.[21] All these are opposite to what you need to do to prepare yourself for a good and relaxed night's sleep.

To deal with this, I find that setting a personal rule that stopping electronic devices at least one hour before bed is an important evening ritual. Arianna Huffington, founder of The Huffington Post and a huge sleep advocate, has a device-free zone around her bedroom and she stops checking her email one hour before bed time to ensure a high quality sleep.[22] While she likes a hot bath before bed, I personally find that taking just a foot bath right before you sleep also greatly improves your sleeping quality as well.

When it comes to stress control, I would highly recommend you try out meditation. You can start out by meditating simply for five minutes each day by just focusing on inhaling and exhaling, while placing your attention either on your breaths or on a point between your eyebrows.

If at any time you catch yourself going off track, just gently bring your attention back to where you set your focus to. A five-minute meditation looks easy and is actually quite hard when put into practice; after staying persistent for a certain amount of time, you will gradually feel the benefits such as obtaining a quieter, more relaxed, and more focused mind, making more conscious decisions, and being more able to withdraw yourself out of a situation instead of becoming too attached to wins and losses, successes and failures.

You can leverage technology to keep yourself accountable. For me, I use a free mobile application called "Insight Timer" to help me time myself and keep track of my meditation journey.[23] You can also explore guided meditation sessions provided in this app to see if such things are your cup of tea.

Here is the catch: there are many meditation practices out there; ironically, too many choices of meditation methods can lead to anxiety in choosing one that works for you or prevents you from going deep into practice. The solution is this: ultimately, you will find out and stick with one type of meditation that resonates with you the most. By going deep into one form of meditation practice, you will truly reap the benefits as you go further.

It is like digging holes in the ground to find water: there are many places where you can dig holes and discover water, but if you keep digging holes in different places without going deep, you will only stay on the surface level and never be able to find that water.

6. Controlling Your Obsession with Cleanliness

Thanks to the gut-friendly knowledge I have accumulated, I now live a more care-free but mindful life. Maybe you could be the same, too?

As I was writing this, there was half a gallon of antibacterial dishwashing detergent sitting inside one kitchen cabinet underneath the sink. I bought this before the knowledge that I need to control my obsessive cleanliness-fantasy of keeping all dishes not only clean but also germ free. To be honest, even the thought of this lit up my mood.

Even after I have become more knowledgeable about how we need to protect the beneficial microbes inside us, I am still keeping it because I feel that discarding it totally seems like a waste to me, and also because there are times I believe using detergents is necessary such as when I need to clean up the chopping board with which I cut raw meat. Having said that, I definitely have become more aware of my daily use and I skip using it when possible.

When it comes to washing cooking pans and dishes, I only use a sponge without any soap to wash them immediately after cooked foods are transported onto plates. The remaining heat in the cooking pan makes it very easy to wash off grease under running hot or tap water, while I can still keep the pan clean afterwards.

I no longer apply my hand sanitizer multiple times a day, which was my former habit. Similarly, I also no longer obsessively wash my hands with antibacterial soap each time after I pet my dog. When I prepare my morning smoothie and a piece of celery falls off the kitchen counter, I simply pick it up and place straight into the blender without rinsing, as long as I am sure I keep my kitchen space clean and use as few harsh chemical treatments for cleaning as possible.

Plus, here is a big bonus for me: my relationship with my dog has become more intimate, which means more hugging, more kisses on his cheeks, and petting.

I am not saying you need to abandon antibacterial soap or other cleaning products all at once. After all, there are times when they come in handy, but what I am saying here is that we all need to start becoming more mindful and consciously evaluating the trade-offs between killing potential pathogens at home and threatening good microbes on your skin and in your body.

When you think using cleaners is necessary, I recommend you opt for healthier products that contain as few questionable chemicals as possible. As a reference, you can go to Environment Working Group's "Guide to Healthy Cleaning" and see what to consider and what to avoid (visit www.ewg.org/guides/cleaners).

My writing and knowledge thus far has got me pondering whether I should apply chemical tick prevention treatment to my dog. Yes, it may be able to prevent ticks; but that may come with side effects like affecting the long-term health of the animal.[24]

Can the application threaten the dog's microbiome and weaken the immune system? And can our contact with our dog indirectly kill our own microbes, too? Current studies on how chemical treatments like tick and flee control on pets are affecting microbes in and on pets and humans are hard to find. This is not surprising at all as there are still a lot of unanswered and unexplored territories in the whole field of microbiota, but since the topic of overuse of antibiotics and excessive cleanliness has been brought up as a potential concern, it becomes a natural problem to me that chemicals used on animals can be an issue for both humans and pets, especially a pet I kiss and hug many times a day.

Before you think that I have been brainwashed during my study of human microbiota, please note that I don't think we have to be too obsessed or too overly stressed out about this. Chemical tick and flee control on animals are still being used; and it seems that many pets and pet owners are doing just fine. Yet we should more intuitively and consciously develop the awareness of evaluating pros and cons of using chemicals on the pets we live with every day and make more responsible decisions.

7. And, Finally... Diversity Matters

As there are tens of thousands of different species of beneficial bacteria inside the body, introducing diversified potential microbial sources increases the chance of your internal microbial community meeting the most ideal environment to help it thrive. This means diversifying your life exposure by participating in activities such as indoor and outdoor gardening, visiting local organic farms, playing with healthy and happy pets, looking into different kinds of food sources for dietary fiber while not limiting yourself to just your one or two most often go-to choices, and varying the types of fermented foods you are introducing into the body.

Although later on this book helps teach you how to get started on this journey with vegetable fermentation, you should always keep in mind that once you master one type of vegetable fermentation, it is important to also try out other types of fermented vegetables and other fermented food groups to diversify your microbial exposure.

Chapter Recap

- *Ways that we are threatening our microbial allies include: caesarean delivery, lack of breast feeding, overloaded processed* foods, overuse of antibiotics, chronic stress, lack of sleep, and obsessive cleanliness.

- *Ways which can help develop a robust microbial community include: a mother vaginally giving birth to her baby, breast feeding the baby, eating more dietary fiber, consuming properly made fermented foods or trusted probiotics supplements, reading food labels as you shop for groceries to avoid processed foods with microbe-killing ingredients, minimizing the use of antibiotics, sleeping seven or eight hours every day, meditation to de-stress, controlling your obsession for cleanliness, and diversifying your sources of exposure to friendly microbes.*

Exercise

- *The first thing I notice that is killing my friendly microbes is _____; because of this, I plan to _____ to start my journey of cultivating a healthy gut environment.*

Checklist

- *Visit www.tracyhuang.me/fvm-resources to download a list of food sources of dietary fiber.*

Chapter 3: Consuming Fermented Vegetables - A Piece of the Puzzle

"For me, fermentation is a health regimen, a gourmet art, a multicultural adventure, a form of activism, and a spiritual path, all rolled into one."

– SANDOR KATZ
Wild Fermentation

I hope the first two chapters have given you the context of what is going on before I invite fermented foods – particularly fermented vegetables – to step into the limelight.

It is important to realize the role of consuming fermented foods plays in this big picture, so that you don't get lost on your path to improving your overall health and wellbeing: consuming fermented foods is a piece of the puzzle that is plugged into the "TCs" (or, Transient Communities) category to help revive your microbial community.

Now that you have your GPS system set in place and know where you are at right now, let's get moving.

What Is Fermentation?

Yogurt, chocolate, bread, soy sauce, miso soup, beer, wine, ginger ale, cheese, vanilla, coffee, and cheese cakes… These are just a few examples of fermented foods that you encounter in your daily life. The word "fermentation" may sound new to you; and, to some people, it may even sound unpleasant or moldy. The fact of the matter is that the concept of fermentation is not at all new to you, especially when you have already been exposed to a lot of the fermented foods.

Simply put, fermentation "is a metabolic process in which an organism converts a carbohydrate, such as a starch or a sugar, into an alcohol or an acid."[1] Overall, in this chapter, you will learn that fermentation is indeed a pleasant experience: it is safe practice; it nourishes your body; it nourishes your mind and emotions, which means you will be constantly having fun with staying creative about possibilities you can make out of fermentation; it nourishes your spirits, because you feel more connected in the web of relationships between you and your own self and between you and nature.

Fermentation as a Safe Practice

The practice of fermentation is as old as human history. It is believed that, in 7,000 BC, people in the early Neolithic village of *Jiahu* in *Henan* province in China had already discovered a way to make a mixed fermented beverage of rice, honey, and fruit.[2]

As discussed in Chapter 2, the recent breakthrough studies of microbiology have raised the awareness of the importance of fermentation to cultivate microbes inside the gut. Once you are struck awake from a dormant state and start to look around the world, you will see that many other countries have been fermenting and even keeping it as a tradition and a national pride for a long time:

To start, sauerkraut, made of white cabbage is very popular in Germany, Switzerland, Austria, the Czech Republic, Slovakia, and Hungary; French people come up with their own version of sauerkraut called choucroute; Lassi, a blend of yogurt, water, spices, and sometime fruits, is popular in India, Pakistan, Sri Lanka, and Bangladesh; Bulgarians loves their kefir, a type of fermented milk; miso and natto (two types of fermented soybeans) are two most loved delicacies in Japan; and, of course, who does not know kimchi, a type of fermented vegetable usually made from Napa cabbage, as a symbol of South Korea?

In China, there are too many fermented dishes to list in here. But here are two interesting examples: one is stinky tofu which is fermented tofu with a strong odor (I know, this might not sound appetizing to you); yet, it tastes surprisingly good and is nutritious for the body. The other one is fermented sticky rice, which is known for warming up the body and replenishing the blood. In China, a lot of new moms consume fermented sticky rice to produce more milk for breastfeeding as well. Interestingly, a friend of mine from Kenya also told me that new moms in Africa use a similar method to produce more milk for infants. What they use is fermented millet porridge.

The ancestors performed fermentation on foods and beverages thousands of years ago; and so your "neighbors" from the global village called the earth have been enjoying it for a long while by now. From the rich history of fermentation and its popularity with people around the word, you can infer that this is a rather safe practice.

Let's explore deeper into the "safety" aspect of fermentation:

First, fermentation is actually used as an effective strategy for securing food safety. In fermented vegetables, rapid proliferation of acidifying bacteria creates lactic acids, acetic acids and other substances like hydrogen peroxide, bacteriocins, and other antibacterial compounds, which builds up an unfavorable condition for pathogenic organisms to establish and grow. Sandor Katz, author of The *Art of Fermentation*, mentions in his book that even when raw ingredients are contaminated, the acidifying environment will inhibit the growth of contaminating bacteria like *Salmonella, Escherichia coli* (or, E. coli), *Listeria, Clostridium*, and other foodborne pathogens.[3]

Because fermentation can effectively ensure food safety, it is used as a method of preservation, particularly for those in cold climates. In those places, fermentation is crucial for survival. As a way to prepare foods to survive famines and severely cold winters, people would catch fish and birds, bury them into the pits, and ferment them for months until it's time for use.

As a side note, now that we are on the topic of food preservation, it is also important to remember this: while fermentation can be used as an effective food preservation technique, you should also caution that not all ferments (fermented foods) can equally preserve foods. For example, it is better to preserve wheat in its dry form; and tempeh's shelf life is only a few days, even when you put it in the fridge.

Next, fermentation can ensure your safety by detoxifying pesticides. A study in 2009 shows kimchi fermentation has a direct impact on the degradation of the organophosphorus insecticide chlorphyrifos. This pesticide was degraded rapidly until day 3 and degraded completely by day 9, thanks to four lactic acid bacteria including *Leuconostoc mesenteroides, Lactobacillus brevis, Lactobacillus plantarum*, and *Lactobaccilus sakei*.[4]

Fermentation can also remove a variety of toxins from foods and the body and, in some cases, turn anti-nutrients into nutrients. When the second atomic bomb was dropped in Nagasaki on August 9th, 1945, physician Tatuiciro Akrizuki and 20 employees were taking care of 70 tuberculosis patients at "Uragami Daiichi Hospital" about 1.4 KM (about 0.87 miles) away from the hypocenter. They did not get any major radiation disease. Dr. Akizuki believed this was due to the consumption of cups of wakame miso soup every day.[5] Putting the story aside, we now have research that shows how miso with a longer fermentation time, can effectively fight against radiation.[6]

Interestingly enough, fermentation can also be used as a strategy to make contaminated drinking water clean and safe, if you add fermented sugars and allow a small accumulation of alcohol and/or acids to destroy the bacterial contaminants. In addition, according to Katz, phytates found in all grains, legumes, seeds, and nuts make these foods unavailable for our absorption by binding minerals. During the fermentation process, the enzyme phytase releases minerals from the phytate bond, thus making it easier for human bodies to absorb the nutrients.[7]

As you see, fermentation is actually a very safe practice. By now, I hope that you have eased your worried mind a little bit. And so let's continue on our journey:

Fermentation for Nourishing the Body

By fermenting different foods, you are able to break down nutrients from these foods into more accessible and bioavailable forms. According to *The Art of Fermentation*, "organic compounds are metabolized into more elemental forms. Minerals become more bioavailable, and certain difficult-to-digest compounds are broken down. In the varied soy ferments, fungi and bacteria digest the bean's prodigious protein into amino acids we can more readily assimilate. With milk, LAB (lactic acid bacteria) convert lactose into lactic acid. Meat and fish are tenderized by the enzymatic digestion of fermentation."[8]

What about grains? A study in 2012 shows that fermentation with *Lactobacillus acidophilus*, *Lactobacillus johnsonii*, and *Lactobacillus reuteri* improved bioavailability of dietary phenolic acids – antioxidants that prevent cellular damage and promote anti-inflammatory conditions in your body, and is found in whole grain barley and oat groat.[9]

Not only are fermented foods able to break down nutrients into more bioavailable forms for your body to get access to, they also enhance the level of nutrients. For example, it is believed that many ferments accumulate increased levels of B vitamins, including thiamin (B1), riboflavin (B2), and niacin (B3), as compared with the raw ingredients prior to fermentation.

More interestingly, fermented foods even create unique micronutrients not present in the raw ingredients. As an example, the

Japanese soy ferment natto contains an enzyme called nattokiase which manages a wide range of diseases such as hypertension, atherosclerosis, coronary artery disease, stroke, peripheral vascular disease, and even Alzheimer's disease.

These are a few examples listed by Katz in his book *The Art of Fermentation*, where he says, "who knows what other compounds as yet unrecognized by science may be present in all our varied ferments?"[10] I totally agree with him here. The scientific exploration of fermentation is just at its beginning stage, while the practice has been in existence since human history and people have been enjoying fermented foods for thousands of years. I have no doubt that there are only more breakthrough discoveries to come related to healing benefits of fermented foods.

Fermentation for Nourishing Your Mind and Emotions

People do not always ferment for practical reasons like food preservation or the mere scientific pursuit on how fermentation is good for the body. Sometimes, it is simply the taste-bud-stimulating flavors that seize people's hearts.

In tropical areas like Sudan, fermentation is used to guide microbial transformations of foods to create different delicacies rather than decomposition. There are so many color combinations and mixed flavors you can make with fermentation: white from shredded white cabbage, green from cucumbers, red from carrots, yellow from whole grains, and purple from eggplants; additionally, you will get sweetness from honey, saltiness from sea salt, sourness from lactic acids, and pungency from spices.

Within the whole experience, you create your own "invention" in the lab called the kitchen. You may do this alone or for your whole family or other loved ones, or you may do this with a few friends or within a local community. The expectation of what will come out, the love you express for the people you are making fermented foods for, and the companion you spend quality time with all add up to a heartfelt experience full of curiosity, hope, joy, happiness, and gratitude. That's why I see fermentation as a way to nourish both your mind and emotions.

Fermentation for Nourishing Your Spirits

You've now learned that fermentation is a safe practice; it not only nourishes the body, but also your mind and emotions as well. But, how is this traditional practice also connected with your spirit?

The breakthrough discoveries that "we are only ten percent human" and that the body is occupied by 100 trillion microbial cells versus "only" 10 trillion human cells (see Chapter 2) makes me feel very small and humble. It is never only about ourselves; rather, we live in a community where we as humans are only a very small fraction in the world called nature.

Fermentation is one of the ways to recognize the existence and contribution of microbes to achieve common good. On one hand, we serve them to help them build a vibrant community within us; on the other hand, they serve us, so that we become strong, healthy, and happy. In this process, we are developing our spirit of serving and recognizing that it is not just about "me". Below is a quote I love from Lisa Heldka – Professor of Philosophy from Gustavus Adolphus College – taken from the book *The Art of Fermentation*:

> *"Fermented foods manifest one of my fundamental philosophical beliefs perhaps better than anything else in my life; in no other context am I so aware of the intricately participatory nature of reality; of the unpredictable interconnections between me and not-me."*[1]

Fermentation provokes our thinking over the relationship between "me" and "not-me". That is very spiritual, deep, and intriguing.

Fermentation and Health

At the beginning of this book, I mentioned that I designed a survey and reached out to more than 100 people to ask them about their top health concerns, and the top five concerns were weight gain, fatigue, acne, digestion, and allergies. After my research on fermentation and the health benefits of fermented foods, I found that these concerns could actually all be addressed by the introduction of fermented foods:

Kimchi combined with a fiber-rich diet can cultivate *Bacteroidetes*,[12] a group of organisms associated with metabolism and the prevention of weight gain; Donna Swchenk told me in an interview that: consuming sauerkraut which contains a high amount of vitamin C, vitamin B- rich kefir, and kombucha strengthens the adrenal glands and relieves fatigue; a mixture of *Lactobacillus acidophilus* cultures and *Lactobacillus bulgaricus*, which you could find from Swiss cheese, yogurt and kefir, can help heal acne;[13] *Lactobacteria* and *Bifidobacteria*, which you can find in a wide range of fermented vegetables, effectively improves digestion; lastly, *Lactobacillus plantarum*, which dominates later stages of vegetable fermentation can effectively treat pollen allergies.[14]

It also amazes me that it is scientifically proven that these types of foods actually address most of the health concerns mentioned. Despite the small sample size, an important message I took away from this survey is that: it seems that there are imbalanced microbial profiles in people in our modern life. This may be why modern people tend to develop health problems that are curable by reinstalling balance in microbial community in human bodies. In other words, by adjusting your inner microbial environments, your overall health is very likely to be improved.

Besides the above five top concerns, here is a short list of health issues that can be healed by the consumption of fermented foods, with potential fermented dietary choices in the parentheses:

- Premature aging and inflammation (fermented milk, fermented bean paste, and fermented mung beans)[15]

- Hormonal issues with women (fermented soy milk which has a stronger effect than unfermented soy milk on reducing symptoms like hot flashes, sweating, shoulder stiffness, arthralgia, and palpitation during menopause, and fermented red ginger which prevents diabetes in postmenopausal women by regulating internal hormones)[16]

- Colds and flu (kefir, fermented vegetables, yogurt fermented with a group of microbes called *Lactobacillus bulgaricus*)[17]
- Crohn's disease (*Lactobacillus casei* and *Lactobacillus bulgaricus* probiotics; you may try *Lactobacillus casei*-containing yogurt, cheese, and fermented green olives and *Lactobacillus bulgariuc*-containing yogurt and Swiss cheese)[18]
- Depression, stress, emotional disorders (*Lactobacillus helveticus* and *Bididobacterium longum* probiotics; you may try *Lactobacillus helveticus*-containing cheddar, Parmesan, and Swiss cheese, and *Bifidobacterium longum*-containing yogurt, fermented dairy foods, and fermented vegetables)[19]
- Diabetes (low-fat fermented dairy products)[20]
- Food allergies and intolerance (fermented milks and milk products recommended for lactose intolerant individuals)[21]
- High blood pressure (fermented dairy products like yogurt, cheese, and sour milk, and fermented soy milk)[22]
- High cholesterol (kimchi and fermented milk)[23]
- Irritable Bowel Syndrome (fermented milk and soy-fermented products)[24]
- Poor memory (fermented ginseng)[25]
- Arthritis (fermented milk, fermented vegetables that contain beta-carotene such as fermented squash, fermented lettuce, and fermented carrots)[26]
- Dry skin (fermented barley and soybean)[27]

During my research, I found that just because a particular strain delivers particular health benefits, it does not necessarily mean foods which contain that strain will deliver the same health benefits to the same degree. For example, just because *Lactobacillus casei* and *Lactobacillus bulgaricus* are proven to have the potential to heal Crohn's disease, does it mean *Lactobacillus casei*-containing yogurt, cheese, and fermented green olives and *Lactobacillus bulgariuc*-containing yogurt and Swiss cheese can deliver the same promise? This may still be an area that is yet to be explored. Despite this, I hope the examples given above are enough to show you the undeniable health effects that fermented foods can give you.

You can go to Chapter 1 to review how microbes are having a positive impact on your different internal systems. My conclusion is that people in the modern days are suffering from microbial imbalances and that addressing gut issues can improve overall health. This idea was later confirmed, when I found Dr. Natasha Campbell-McBride online, a formerly trained Russian neurologist and founder of the Gut and Physiology Syndrome (GAPS) Nutritional Program for the treatment of autism and a wide range of neurological, psychological, and autoimmune disorders. She specifically points out on her website that a total of *86* health issues can be healed by fixing the gut.[28]

Fermented vegetables can even help with small intestinal bacterial overgrowth (SIBO) and acid reflux. These are health concerns that may cause you to worry whether introduction of more acid and bacteria into the body – in this case, it is consuming fermented vegetables – will worsen your health conditions. Interestingly, consuming acidic fermented vegetables can actually relieve such conditions. I will cover this in more detail in Chapter 4.

As a side note, although fermented foods present such potential in solving various health problems, I don't believe they are a panacea. But the healing power behind this type of food certainly is worth your attention to at least give it a try for the sake of your own health.

What Foods Can Be Fermented?

Almost any type of foods you can think of can be fermented. Browse different recipe books regarding fermented dishes on the internet for a while and you will soon be able to find cooking instructions for fermenting meat, fish, eggs, vegetables, fruits, grains, dairy, bean, legumes, nuts, seeds, and sugars.

You don't want to limit your options when it comes to fermentation. However, at the same time, if you are a newbie to the fermentation world, you also don't want to overwhelm yourself by having too many options, not knowing where to get started (analysis paralysis, anyone?). That's why it is wise to just stick with one group as you get started.

So, which group shall you pick first? Since this book is primarily about fermented vegetables, I guess you know the answer already. I'll go deeper to explain why.

Why Start with Fermented Vegetables?

While there is no right or wrong answer to pick which food group to kick off your fermentation journey, you cannot go wrong if you make sauerkraut or other fermented vegetables first. As Katz says, "Fermenting vegetables is the ideal way to begin a fermentation practice in your life."[29]

Additionally, it involves low risks. "As far as I know, there has never been a documented case of foodborne illness from properly fermented vegetables," says Breidt (the microbiologist who specializes in vegetable fermentation).

To learn anything, you need to achieve mini-milestones and get a quick win soon after you get started. This way you can sustain your interest and build up the momentum in learning a new thing by celebrating your little victory. Fermenting vegetables allows you to do so, because it is very easy to prepare raw ingredients and make your first fermented vegetables. For example, it can take you as few as five minutes (or less) to prepare ingredients for fermented beet juice and enjoy beet kvass in just four days.

The last but probably the most important reason I think you should get started with making fermented vegetables relates to your gut health. Your gut health is threatened in modern days, which should call for your immediate attention: the beneficial intestinal bacteria are reduced due to the frequent exposure to antibacterial chemicals and other conditions as mentioned in Chapter 2. By replenishing the lactic acid populations within you, you not only restore the balance between "good" and "bad" bacteria, but also cultivate and protect lactic acid bacteria, one of the most important types of bacteria found in the digestive system.

How Are Vegetables Fermented?

I see that there are altogether four phases involved in this vegetable fermentation process:

First, when vegetables are not harvested, lactic acid bacteria (LAB), most commonly *Leuconostoc mensenteroides*, stand for only one percent of a plant's microbial populations.

As soon as the plant is harvested, more nutrients become available from the cellular contents of ruptured tissue, leading to an increase of the total number of microorganisms, and another round of distribution among different types of microbes starts to take place. Aerobic bacteria dominant on the living plant are replaced by anaerobes such as LAB, hence the drastic increase of *L. mensenteroides*.

When vegetables are submerged, *L. mensenteroides* start initiating fermentation. This type of bacteria is heterofermentative, which means the fermentation process generates the primary product, lactic acid (hence the sour flavor), and a significant amount of carbon dioxide, alcohol, and acetic acid. That's why you will notice bubbles in the early stages of vegetable fermentation.

As the fermentation process develops, the environment becomes more acidic. This leads to a shift to the acid-tolerant population like *Lactobacillus plantarum*. This kind of microbes is called homofermentative bacteria (in this case, it means the fermentation process will generate primarily lactic acid) and characterizes later stages of vegetable fermentation.[30]

What Vegetables Can Be Fermented?

Theoretically, all vegetables can be fermented, but not all types can come out with the same texture or be fermented equally well. Below is a quick summary of Katz's recommendations:

The most popular option are cold-weather vegetables which includes cabbages, radishes,, beets, broccoli, carrots, English peas, kohlradi, lettuce, onions, and potatoes. Roots vegetables – such as radishes, carrots, turnips, beets, rutabaga, celery roots, parsley roots, okra, and earthy burdock – are also ideal. Peppers of all kinds can ferment very well; as can eggplants, chard stalks, beans, and vegetables from the Brassica family such as Brussels sprouts and cauliflowers.[31]

I learned from my experiments that the tastes of radishes, carrots, white cabbage, red cabbage, Napa cabbage, beets, bok choy, eggplants, and cucumbers after fermentation were very good. I personally really enjoyed the crunchiness of these vegetables, if fermented with the right techniques (yes, even eggplants can taste a bit crunchy after being fermented). If you are interested in how I made them, you can refer to Chapter 12 for the recipes and my lessons learned.

According to Katz, you should pay attention to cucumber and summer squash fermentations, as these two vegetables can get mushy easily; it is recommended you eat them quickly when you are done with fermenting rather than aging them. Luckily, I also found that, with the addition of perennial leaves such as black tea leaves and grape leaves (I used the former), cucumber pickles could taste crunchier and remain so for quite a long time after transported into the fridge. Make sure that you add a little of this type of leaf as you make cucumber pickles if you enjoy crunchy textures. You will learn more on how to make fermented vegetables in Chapter 5.

It is also good to know up front that dark leafy greens like kale and collards which are rich in chlorophyll will develop "a very strong characteristic flavor" during fermentation, which may not appeal to everybody.[32] Curious about this, I specially did an experiment of fermenting baby kale. I was glad that I knew this piece of knowledge in the back of my mind beforehand; so, it didn't turn out to be very surprising that the fermented baby kale didn't smell or taste very pleasant to me personally.

At this point, you have learned the big picture that Section 1 has drawn for you: in Chapter 1, you have learned that the indispensable yet often-neglected warriors called microbes – particularly, gut floras – are guarding you tirelessly and require your immediate recognition and respect; in Chapter 2, I have shared that there are areas you can control to minimize the chance of killing these allies and ways you can cultivate them and help them thrive to better serve you; and, finally, in this chapter, you have learned that it is very likely that your overall health will improve as you begin to make and consume fermented vegetables.

Now it's the time to move onto the next section, where I will guide you to become an expert in vegetable fermentation in six steps, beginning with removing your mental blocks which may prevent you from trying out fermented foods. I will soon address 16 areas where you may have misconceptions or concerns. Are you ready?

Chapter Recap

- *Fermentation is a safe practice because it is as old as human history; has been an international activity since ancient times; can effectively ensure food safety; is a method of preservation; detoxifies pesticides; detoxifies the food and the body; and turns anti-nutrients to nutrients.*

- *Fermented foods nourish your body by breaking down nutrients into more accessible and bioavailable forms, by enhancing the level of nutrients, and by creating micronutrients not present in the raw ingredients.*

- *Fermenting foods nourishes emotions because it can be a fun and adventurous experience. You get to taste different flavors by either following established recipes or creating your own; the whole experience is also heart-nourishing as you do it with all kinds of positive emotions such as curiosity, hope, joy, happiness, and love.*

- *Fermentation also nourishes your spirits, as the practice reminds us all that "we are only ten percent human". It invites us to go beyond "me" (or, ego) and see ourselves living in a larger community where 100 trillion microbial cells also exist and where we should serve them as well to co-create a healthy system.*

- *A myriad of health problems can be addressed by fermented foods including, but not limited to: overweight, fatigue, acne, digestion, allergies, aging, hormonal issues with women, colds and flu, Crohn's disease, depression and stress, diabetes, food allergies and intolerances, high blood pressure, high cholesterol, side effects from taking antibiotics, emotional eating, lack of* focus, poor memory, inflammation, arthritis, hyperactive thyroid, and dry skin.

- *It seems that there is an imbalanced microbial profile in people of the modern age. By adjusting your inner microbial environments, your overall health is very likely to be improved.*
- *Almost any type of foods can be fermented; vegetable fermentation is the ideal way to start because it involves low risks, is easy to make, and replenishing lactic acid bacteria is crucial for your gut health*
- *Vegetable fermentation starts with hetero-fermentation, which means that carbon dioxide, alcohol, and acetic acid will be produced; as the environment gets more acidic, homo-fermentation takes place, when lactic acid predominates.*
- *Theoretically, all vegetables can be fermented. The preferred vegetables are: cold-weather vegetables, root vegetables, peppers, eggplants, chard stalks, beans, vegetables from the brassica family. Dark leafy greens, when fermented, can leave a strong smell and flavor that may not appeal to everybody. My vegetable fermentation experiments including the recipes and lessons learned can be found in Chapter 12.*

Exercise

- *I have learned that by adjusting my _____, my overall health is very likely to be improved; and, I can get started with _____ because this practice is safe and easy and produces lactic acid that is crucial for my gut health.*

SECTION II: MASTER VEGETABLE FERMENTATION IN SIX STEPS

Chapter 4: Step One – Demystify Your Top Concerns

"Fermenting vegetables is a simple, inexpensive process that was used reliably for a few thousand years."

– KIRSTEN SHOCKEY & CHRISTOPHER SHOCKEY
Fermented Vegetables

When I reached out to 100 people for my survey, I found out that people were as interested in finding out answers to their own worries and concerns regarding consuming fermented foods as in learning the health benefits of this type of food. For some people, they were even more interested in the former. Therefore, instead of directly guiding you on how to make fermented vegetables, I find it necessary to address existing questions, doubts and worries you may have.

Below are 16 areas that will help you understand more about fermented foods from a safety perspective.

1. Aren't acidic foods bad for one's health?

Okay, so you are assuming two things: first, "acids are bad for the body;" and, second, "fermented foods are all acidic." But, these two assumptions are not true. I am going to address your first concern here, while I will get to the second assumption in my answer to the second question.

The meals that shape our diets these days are overly acidic such as meat, coffee, wine, beer, cheesecakes, French fries, fried foods, hamburgers, ice-creams, popcorns, pizzas, and milkshakes, the excessive consumption of which will gradually lead to an overly-acidic and sick body.[1]

It is the IMBALANCE of the pH in the body - not ACIDITY inside the body that causes sickness. As a matter of fact, if the body gets too alkaline, your body can become sick, too.[2]

This helps you understand the relationship between acidity, alkalinity, and optimal states of health and wellness: it is not that acidity means bad or alkalinity means good (in fact, we need both at the same time and should not have excess amount of both); it is the imbalance between acidic and alkaline levels inside the body that causes sickness.

Under a healthy condition, your blood manages to maintain a pH level between 7.35 and 7.45, which is slightly more alkaline than pure water (pH of pure water is 7, which means neutral).[3] By modifying pH levels inside the body while keeping a certain level of acidity, you adjust the body pH to slightly alkaline to stay in good health.

Now that you have a better understanding of the acidic foods, let's have a look at the health benefits of some of the acidic foods – such as whole grains and mixed nuts, if you consume them at a controlled amount.

The two-thousand-year-old Traditional Chinese Medicine categorizes "whole grains" to the "Earth" element, which means the foundation of life; and believes that one should consume a certain amount of whole grains like brown rice, oats, millet, and barley daily to nourish the body (particularly the stomach and the spleen) and to prolong life.

As another example, peanut butter contains fiber, vitamins, and minerals; and consuming it frequently can lower the risk of having heart disease and type 2 diabetes.[4]

As you see, it is not that acidic foods are bad, because they (mostly whole foods) have their own nutritional value and significant health benefits, too. It is the imbalance that is causing problems. But of course, I would recommend minimizing the intake of acidic processed foods like French fries, soda drinks, and packaged fried potatoes chips because they have little nutritional value while bringing a lot empty calories to your body.

Now, you are also assuming all fermenting foods are acidic. Yes, they are acidic; yet, some of them are alkaline-forming. What does that mean? Let's move on to learn more.

2. Some people say that cultured vegetables are alkaline; and some say they are acidic. Which is true?

Before we dive into the question, we need to define the difference between "acidic" versus "acid-forming", and "alkaline" versus "alkaline-forming". "Acidic" and "alkaline" refer to the food property itself before a particular food is consumed, whereas "acid-forming" and "alkaline-forming" are used to describe pH environment *after* the type of food is digested by the body. As you see, using terms "acid-forming" and "alkaline-forming" are actually more practical to measure the impact a specific kind of food has on the body.

For example, a lemon is acidic by nature. But after lemon juice is consumed and digested, it leaves alkaline residues inside the body and therefore actually alkalizes the body. That's why we say lemons are alkaline-forming.

Fermented vegetables are acidic; yet it doesn't mean that they are all acidic-forming. As you may know by now, fermentation makes nutrients in food more bioavailable. Since vegetables are a very good source of minerals, which are alkalizing for the body, fermented vegetables can be alkalinizing to the body, too. You can find traces of evidence here and there.

For example, Katz argues that *"Bacillus subtilis* var. *natto* (formerly known as *Bacillus natto*), the bacteria that transform soybeans into natto, are alkaline-forming rather than acidifying." Miso soup and sauerkraut are alkaline forming, too.[5]

Granted we cannot make a conclusion that fermented vegetables are all alkaline forming just because natto, miso soup, and sauerkraut are alkaline forming. In fact, there are even mixed opinions that miso soup and sauerkraut are acid-forming. Despite the controversy around the alkaline-forming and acid-forming properties of fermented vegetables, I still consider consuming fermented vegetables to be on the safe side.

Personally, I sense that the deeper level of worry here is you are wondering if fermented vegetables are healthy for the body. It is important to remember that alkalinity and acidity is not the only benchmark to decide whether a particular type of foods is healthy or not; rather, we should all look at a broader picture: we should recognize how the study of microbiology is evolving (as described in Chapter 1) and how fermented foods are playing a role in this process; how LAB from fermented vegetables are benefiting the body (as described in Chapter 3); and how different people around the world are sharing stories on how fermented foods are healing themselves and improving the health of their families.

Whether fermented vegetables are tested out to be alkaline or acidic, we should also remember that one is always recommended to consume a small portion of fermented vegetables every day (more can be found in Chapter 7). So the bigger picture here is how you need to keep yourself alkaline overall by having a well-balanced plant-based diet, instead of focusing your entire energy on worrying about whether fermented food is alkaline or acidic.

3. How can I make healthy and low-acidic fermented vegetables?

Let's pause and go deeper into what prompts you to opt for low-acidic fermented vegetables. There are three potential areas that you might be concerned about at this point.

Is it that you are afraid that acidic foods are bad for health? If so, then you can refer to my answer to question #1. Are you are currently experiencing acid reflux and afraid that you might not be able to take in more acid? If so, you might want to check out my answer to question #4 that immediately follows. Or, does this question simply reflect your personal flavoring preference to less sour-tasting foods which are more appealing to your taste bud? If so, here is the general tip: the longer you wait, the sourer it gets. But at the same time, you don't want to start eating fermented vegetables too soon just to avoid excess sourness.

Instead, you should make sure that the pH reaches a certain level before consuming them.

pH 4.6 is the key cutoff, according to Breidt, because that is the point when *Clostridium botulinum* gets killed. This bacterium can cause botulism, which is a rare but serious paralytic illness which leads to weakness in the body, poor vision, trouble speaking, and even potentially death.

4. Are fermented foods good for people suffering from acid reflux?

You may think that there is too much acid inside your body already, why should you introduce more "trouble" to the body? Contrary to what you think, acid reflux is actually a sign of having too little stomach acid inside the body, hence poor digestion. Sauerkraut and cabbage juice can help address this problem to stimulate more stomach acid and improve digestion.[6]

At the same time, too little stomach acid and poor digestion can lead to candida overgrowth. If you are suffering from candida overgrowth, then at this stage it is wise to cut out all grains, starches, sugars, and any foods that feed the candida cycle. Again, cultured foods and beverages will support the development of a healthy intestinal flora, help break this candida cycle, and supply enzymes to improve digestion.[7]

With all this in mind, you should also remember that: despite all of the benefits fermented foods may have for acid reflux, this doesn't mean that they are enough for healing acid reflux completely. After all, they are not a panacea. Of course, it's always recommended that you consult your doctor before making a final decision.

5. Sorry, but I am afraid that I don't like the taste and flavor of fermented vegetables.

Did you make this assumption without even trying them first? Did you know that you can train your gut to like those foods down the road by consuming a tiny amount at the very beginning and gradually increasing the amount? When I first made and tried fermented beets and beet juice, they tasted strange to me; and I was not used to the tastes at first, as the whole concept of fermented juice and any other fermented vegetables other than cucumber pickles, sauerkraut, and kimchi was new to me. But since fermented beet juice, or beet kvass, is highly praised by a lot of fermentation experts out there, I thought to myself: "this can't go wrong." And then I started with one tablespoon of juice. As I gradually got used to the flavor, I began to enjoy it and even came up with different ways to prepare it, such as mixing beet kvass with vegetable and fruit juices.

My lesson learned here is that we need to differentiate between "not being used to a taste" and "not liking a taste", and be open-minded to unexplored territories. So give your gut some time to adjust and get used to the new environments you are about to be introduced to.

If you cannot handle the level of sourness from fermented vegetables, there are ways to trick your taste-buds by mixing them with other types of foods to moderate the sour stimulation to your tongue, like mixing sauerkraut with unsalted quinoa for breakfast.

The bottom line is: do not let your emotional drives (worry and dislike) and sensational direct feedback (the thought of "I don't think I will like the taste") stop you from trying something that your body yearns for (good microbes) and cause to you give up trying even when you have never given it a chance. Once you understand the big picture – that healing the gut is what modern people need to do to improve their overall health and wellness – you will then need to balance your internal involuntary feelings and decide on the better action for your body and long-term health.

As a matter of fact, a lot of the fermented vegetables, if prepared well and properly, not only smell pleasant and taste good, but are fun to make, too.

So, do not let you negative presumptions and mental associations prevent you from trying something. As always, look at the big picture.

6. Can they negatively affect people with small intestinal bacterial overgrowth (SIBO)?

Based on the limited amount of research so far, whether or not probiotics can treat SIBO is a controversial topic, as there are mixed study results regarding this. This means: despite the controversy surrounding consuming fermented foods to heal SIBO, one cannot *completely* ignore the potential contribution this type of food can give to your health. The fact that a Specific Carbohydrate Diet (short for SCD) and Gut and Psychology Syndrome diet (short for GAPS diet), both of which are considered effective diets for SIBO, advocate the use of probiotic foods and pills also gives credibility to fermented foods.[8]

The causes of SIBO are very different, such as lack of stomach acid, damage to the intestine by toxins like alcohol, diabetes, surgical removal of the small intestine, etc.[9] Therefore, the question of whether probiotics can help treat SIBO is often too broad and generalized. By going deeper into a health problem, you may have a better chance of solving it more effectively. For example, if SIBO is caused by low stomach acid levels, then consuming fermented products creates a good chance of healing you, because they can actually help relieve low stomach acid problems and improve digestion.[10] So, you may want to dig deep into your SIBO problems.

To learn more about SIBO and whether fermented foods could play a role, I reached out to Donna Swchenk. According to her, the base of the problem is that the body is out of balance. She believes that gut problems could be the base of the issue, and that "healing the gut would go a long way in helping restore the body back to its natural state," as she addresses in her most recent book *Cultured Food for Health*.[11]

7. Why are there many reported health concerns of consuming fermented vegetables? Like, fermented foods causing cancer?

The ratio matters. There is a balance in everything. If you consume too many fermented vegetables with higher sodium content on a regular basis, there will be a higher risk of different health issues. Oftentimes, it is this excessive amount of intake combined with a factor like high sodium levels that can cause problems.

In fact, it is high levels of sodium intake that can lead to stomach cancer and other health problems.[12] According to Dr. Poh Gek Forkert who specializes in anatomy and cell biology from Queen University, you don't have to be too apprehensive about consuming fermented foods and beverages, because they will not pose a risk if you consume them at low levels.[13] Similarly, *LA Times* reported that kimchi, if not excessively consumed, actually could give the body a lot of health benefits. It is only very high intakes of kimchi that can increase your gastric cancer risk.[14]

The amount per serving should always be moderate. As a general rule of thumb, just because one type of food is proven to bring health benefits, it doesn't mean the more you eat, the more benefits you will get. It is again about keeping the balance. You can find details about how to safely and properly consume fermented vegetables in Chapter 7.

One thing I have gradually learned over time is that we should all use our judgment when facing science, as science can point to different directions and lead you to more confusion. It is good to keep a skeptical mind when it comes to eating fermented foods, so that we all become savvy customers and eaters.

When I see conflicts during my scientific research, I usually go with industry experts that resonate with me like Dr. Mark Hyman, Dr. Joseph Mercola, and Dr. Andrew Weil.[15] Another good tip is to look at the wisdom of authentic bloggers who share their experiences on how fermented foods have helped them and find a pattern in what they share. For example, here is a piece of advice on Quora, an opinion sharing site, by a "kimchi-driven cultural anthropologist" who gives me a point of reference as I did my research on safety of fermented foods and whose opinion reinforces the idea that it is keeping the balance that matters:

"There are many Korean families who are aware of the dangers involved with high-intake of salty foods, and they balance that out very nicely, but there are also many who don't. Just don't stuff your face with spicy and salty foods."[16]

At the end of the day, everything should be in moderation: moderation in the amount of fermented vegetables and sodium intake. And, don't forget to look to your own body and observe how you look and feel before you conclude what works best for you.

8. Will fermented vegetables have too many nitrates and nitrites?

This is a popular question among Chinese people. Almost every time I talked to the Chinese, the first response I received was the concern over whether excessive amounts of nitrates and nitrites in fermented foods would cause cancer.

In fact, even fresh vegetables you eat – such as lettuce, spinach, parsley, and collard greens – contain nitrates and nitrites. What may be more surprising to you is that they are actually produced by your own body "in even greater amounts than can be obtained from food", according to Chris Kresser, a globally recognized leader in the fields of ancestral health, Paleo nutrition, and functional and integrative medicine. He further explains, "Salivary nitrite account for 70-90% of our total nitrite exposure."[17]

Ingested nitrates are converted to nitrites when in contact with saliva or are secreted in the urine within five hours of ingestion. Some nitrites in the stomach reacts with gastric contents, which forms nitric oxide that may show health benefits. Nitrites also circulate inside the body and may function as antimicrobials in the digestive system to help kill pathogens like salmonella.[18]

When you look from a broader scope, you quickly see that nitrates and nitrites are basically everywhere in foods, and even already existing in the body. Further, a certain amount of them directly contributes to your overall health and well-being, too. With all being said, it seems that one should not be overly concerned about nitrates and nitrites in foods.

If you are concerned about the safety of consuming fermented vegetables, I then recommend you use a pH test meter to keep track of the pH levels of the ferments you make. "Generally, the longer you let it go, the lower pH you get, (and) the safer you will get," says Breidt.

So far, research is still ongoing to look deeper into this area. But as long as you follow proper instructions of making fermented vegetables and consume them in moderation, it is still considered to be safe.

You can find details of testing the pH in ferments in Chapter 5, making fermented vegetables in Chapter 6, as well as how to consume fermented vegetables safely in Chapter 7.

9. I am afraid of bacterial contamination or moldy products.

Fermenting foods doesn't necessarily equal getting bacterial contamination or moldy products. If you follow proper preparation procedure, you should be able to minimize the risk of getting a moldy batch of fermented vegetables. Yeasts and molds develop when fermented vegetables are not properly made by being exposed to air. While you are making fermented vegetables, as long as you make sure that vegetables are suppressed in the brine (salty water) away from the air, you should be in good control.

Even if you see surface yeasts, you don't have to always worry about them. Usually skimming the brine surface will be okay; but if you see there is a thick layer of yeasts and molds on the top of the jar, you should throw away the whole batch, according to Breidt.

One effective way to make sure that the vegetables you ferment are safe to consume is to take on a scientific approach to decide whether they are ready to serve. This is when measuring the pH of the ferments is helpful. The presence of molds and yeasts can bring up the pH. If the results show that the pH reaches 4.6 or lower, that means it is still safe to consume.

10. If consuming too much sodium can cause cancer, won't consuming fermented foods increase my sodium levels and increase my chance of getting cancer?

Don't fixate your attention to just fermented foods. It's important to first take a step back and evaluate what you are eating in general. Even if we stop thinking about fermented foods entirely, chances are that you are already over-consuming sodium: a study in 2012 showed that average sodium consumption is too high.[19]

According to U.S. Food and Drug Administration, people who are aged two and above should have no more than 2,300 mg of sodium per day, which is about one teaspoon of table salt.[20] The following demographics should consume even less salt on a daily basis: people who are over 51, who have black skin, and who suffer from high blood pressure, diabetes, and chronic kidney diseases. These people should have no more than 1,500 mg per day.[21]

There are many ways you take in sodium that you might not have been aware of, and sodium adds up very quickly in your daily diet to more than one teaspoon a day.[22] For now, your sources of sodium may come from: bread and rolls, cold cuts/cured meats, pizza, poultry, soups, sandwiches, cheese, pasta mixed dishes, meat mixed dishes, and savory snacks such as salted peanuts, chips, and pretzels, just to name a few.

Based on this, if you are basically eating processed foods and eating out often, chances are you can easily take in more sodium daily than recommended. Consuming any more fermented foods soaked in the brine will increase the amount of your sodium intake and may bring in potential health problems down the road.

If you want to enjoy fermented vegetables prepared in the brine without getting sodium overload, the first step would be to evaluate your sources of sodium and cut down the amount of processed foods like pasta, spaghetti, pizza, bread, and canned foods. If you find out you already take in too much sodium every day, then consuming salt-free fermented vegetables such as salt-free sauerkraut is a better option for you at this point.

Even if you are currently controlling your sodium intake well, methods that require salt like adding the brine, pickling, and making kimchi can easily mean over-consumption of sodium. So, always make sure that you find the balance and keep your sodium intake in check.

11. Can I not just buy commercially fermented vegetables? Why bother making home-made fermented vegetables?

While it is possible to buy quality commercial fermented products on the market, there are a lot of products you should be cautious about. Should you not stay careful, you may end up getting a fermented product that kills your microbes and causes a series of health issues.

So, what's wrong specifically with commercial fermented foods?

First, very often you will see that vinegar is on the ingredient list, which may suggest they have not gone through a traditional fermentation process. According to Breidt, most pickles bought at the grocery store are "fresh pack", which means that fresh cucumbers are put into acidic brine made of vinegar and spices then pasteurized. These retail cucumber pickles are not fermented; and the vinegar is used as a preservative and flavoring ingredient. Therefore, the presence of lactic acid bacteria is not guaranteed.

Then, there come the problematic additives. When I shopped at a local supermarket and randomly picked up four bottles of fermented vegetables, I found two out of four of them contained high fructose corn syrup (HFCS). You are very familiar with this sweetener by now. It can cause a leaky gut and deprive microbes of the mucus-foods they rely on.

Speaking of HFCS, there are other forms of added sugar out there in fermented products, too. Excess sugar is added to commercial products to appeal to your taste buds, but this practice can also trigger a series of problems like weight gain, increased triglycerides, tooth decay, and heart disease.[23] *The Good Gut* by the Sonnenburgs clearly mentions in their book that "if cane sugar, corn syrup, or any other types of sugary sweetener is listed within the first three ingredients (on the food label), avoid that product."[24]

Another questionable additive is sodium benzoate, which is added as a preservative. The FDA states that it is okay to use if the amount is less than 0.1% by weight.[25] You will have to especially watch out for consumption of ascorbic acid, or vitamin C, because when the two meet, cancer-causing benzene is formed.[26]

But, wait.

Pickles themselves most likely contain vitamin C, so how can it be avoided when you put them in the same jar? In addition, you have to take into consideration what you eat during the day, too. For me, I have my multi-vitamin supplements, which contain ascorbic acid, in the morning before my breakfast. If you take supplements, you will then have to pay attention to how you want to schedule the time to have ascorbic acid-containing supplements and commercial fermented products that contain sodium benzoate separately. To me, this just gives me extra burden and mental stress.

Granted that a very small amount of sodium benzoate or even benzene would not be a problem and I also believe in the body's self-defending and cleansing mechanism,[27] I still recommend you consider alternatives first – such as buying commercial fermented products by health conscious manufacturers or simply making fermented vegetables yourself at home – before you allow this chemical to enter your body.

Now, what are the other concerns? As a solubilizing and dispersing agent, polysorbate 80 found in pickles has been linked with severe allergic reactions.[28] Alum, a firming agent that helps the cell walls of fruits and vegetables become sturdier – so that you can get crunchy pickles – is toxic when used in large amounts (or accumulated from many small consumptions). Alum can cause irritations and damaged nervous system tissues.[29]

The current trend is to reduce reliance on chemicals to improve texture, so if you make your own fermented vegetables, there are safe and natural ways to enhance crispness, such as adding grape leaves or black tea leaves, which you will find more in detail in Chapter 6.

As I am writing this, I am thinking to myself: isn't it ironic that you are eating a packaged product that you think will give you good microbes but in fact will kill them because of the additives from the food processing? If lactic acid alone can serve as an effective preservative, why should we use additional synthetic preservatives?

HFCS, added sugar, sodium benzoate, polysorbate 80, and alum are just a few ingredients I found that can be toxic to the body after I examined only four jars of commercial fermented products; there might be more hidden concerns out there in chemicals we haven't discovered yet.

Another concern is pasteurization. The live cultures in fermented vegetables cannot tolerate heat exceeding around 115 °F (47 °C). As Katz mentions, "many packaged mass-produced ferments are pasteurized for shelf stability, thus destroying the live cultures. To receive the benefits of live cultures, you must obtain these foods unpasteurized, or make them yourself."[30]

Next time if you look closely when you do grocery shopping, you're very likely to find that some fermented products are actually placed on the shelf at room temperature. What's wrong with that? Well, fermented vegetables must be refrigerated to be preserved, as lactic acid bacteria are sensitive to heat; so the chances are that these products may not have living microbes.

The way that most commercial fermented products are processed, the presence of problematic additives, pasteurization, and exposure to room temperature are just a few reasons for why choosing retail fermented product these days can is a concern. All these factors makes me feel that creating my own fermented vegetables is a better choice, as I can control what goes into the jar, the fermentation process with the right temperatures, and the temperature under which I preserve them.

Besides, making fermented vegetables at home is more cost-effective; it gives you the opportunity to turn your kitchen into a food lab and feel the freedom of control and express your creativity by mixing and matching different types of ingredients, adding your ideal amount of salt, and controlling the duration for fermentation. This can make you more emotionally connected with whoever you are making these vegetables for, because you pour your love and efforts into jars as you are fermenting vegetables.

Having said that, if you still decide to buy commercial fermented products, be sure you make reading food labels a religious practice. Once again, you should consciously watch out for excessive added sugar, HFCS, sodium benzoate, polysorbate 80, and alum on the label, and understand that commercial fermented vegetables may not be processed properly to generate the beneficial lactic acid bacteria. If you encounter chemical-looking vocabularies and you aren't sure what they are, get into the habit of Googling them or referring to credible sources.

12. What is the safest way to consume fermented vegetables?

As Breidt points out, fermenting vegetables is the safest way to start your fermentation journey. So, your choice of getting started with vegetable fermentation will be quite safe no matter what. The four aspects below are what I think are crucial to keep in mind to make sure that consuming fermented vegetables remains this safe.

First, consider mold control. As long as you make sure that all ingredients are well submerged underneath the brine, you should be able to avoid mold growth in your batch. Then, it is all about acidity level control. When the pH hits at least 4.6, it means that harmful bacteria are killed off during the fermentation. This is a good benchmark for newbies to have as a reference to ensure a safe experience.

When it is time to consume fermented vegetables, always remember moderation is the key. However you like the flavor or texture of the ferments, you should always consume fermented vegetables in small amounts. Next, it is the sodium intake. With one teaspoon of table salt as the recommended standard for sodium intake each day, it is very easy to go over the suggested amount. So, be mindful of the sodium from all kinds of foods you take in during the day and decide judiciously how much fermented vegetables you want to eat for the day. For more tips on how to eat fermented vegetables properly, you can check out Chapter 7.

13. How safe are fermented vegetables for kids and pregnant women?

Fermented vegetables are not only important for children and pregnant women, but also are "particularly crucial" for them, according to Dr. Mercola.[31]

The introduction of healthy bacteria to kids at an early age can have a profound impact on their health, including brain development and other areas later in life. It is proven that there is a direct connection between kids with autism and damaged gut flora; and supplementing gut flora is a crucial way to help these young children.[32]

In my interview with Donna Schwenk, I learned that children tend to be put on antibiotics at an early age these days, killing the beneficial bacteria inside the gut and thus leading to the tendency to develop allergies and different types of food intolerances. The good news is that fermented vegetables and other types of fermented foods could help.

Since moms are carriers of babies, Swchenk further expressed that the more beneficial bacteria the mom has, the more she is going to give them to her babies, which will become the foundation for the baby's life. Fermented foods including vegetables and other food groups are listed as part of the diet plan for pregnant and nursing women by Weston A. Price.[33] I also found one study from Dr. Campbell-McBride, the founder of GAPS Nutritional Program. According to the study, 100 percent of moms of autistic children have abnormal gut flora and health problems related to it; and the grandmothers of those children also have abnormal gut flora, but at a much milder degree. This suggests that there is a generational build-up of abnormal gut flora, with each generation becoming more prone to being further affected.[34]

To help the next generation and more generations that follow, it all starts with the moms healing their guts, a conclusion Schwenk also touched upon during our interview.

14. Are fermented vegetables gluten free?

Gluten is a protein that naturally occurs in a number of grains such as wheat, triticale, barley, ray, and oats. Since vegetables and grains are two different food groups.[35] This means there is no gluten in vegetables. Therefore, the answer to that question is yes, raw vegetables are naturally gluten free.[36]

15. What is the difference between probiotic supplements and fermented foods?

Both have their own benefits; it is hard to decide which one is better. To get started, here is a deeper understanding of probiotics.

Within a species of bacteria, there are different strains. Probiotics mean specific strains which have been characterized to show specific human health effects, while other strains of the same species may not have them and in fact probably don't. The USDA researcher Breidt mentions that, to solve a particular human problem, we might need very strain-specific probiotics to solve it.

During my interview, he further explained that "there might be general characteristics of lactic acid bacteria that could be helpful to you. But some *Lactobacillus* strains might not be as helpful to you in fighting disease, which requires a specific strain in the species to trigger immune system to reduce incidences of disease, whereas other strains may not."

Similarly, Mary Ellen Sanders, a probiotics consultant and director of the International Scientific Association for Probiotics and Prebiotics, points out that: "The distinction between probiotics and live cultures is important relative to the presence or absence of data validating health effects in humans... A recommendation that patients consuming antibiotics eat yogurt with live cultures is weak compared with a recommendation to consume a specific probiotic product that has been studied in human studies and shown to reduce antibiotic-associated side effects. Untested products still may have an effect, but cannot be recommended strongly."[37]

On the other hand, a probiotic supplement usually contains no more than 10 billion colony-forming units, whereas there can be as many as 10 trillion colony-forming units of bacteria in fermented vegetables. That's why Dr. Mercola has come to a conclusion that "one serving of vegetables was equal to an entire bottle of a high potency probiotic!"[38]

Here is another reference point from *The Good Gut*: the Sonnenburgs state that it is hard to predict what specific effect a particular group of probiotic bacteria can have on a human body, because everyone's microbiota is highly customized. This means one probiotic supplement can work for your friend, but may not work for you. That's why they believe that consuming fermented foods seems to offer a better chance for encouraging a microbe that will have a positive effect on a human body due to their attainment of a diverse collection of microorganisms.[39]

They also point out the loose regulation in the current probiotics market. Currently, manufacturers do not need FDA approval to put probiotics on the market; and live microbes attained are often not in line with what has been written on the label. The solution is to go to trustworthy companies that provide information and studies with their products, and that have labels with the types of bacteria and the shelf life.[40]

Despite the different voices, there is one thing that's for sure – the benefits and positive influences by both probiotics and fermented vegetables and other food types should be recognized.

It seems that when we are dealing with a very specific health problem, probiotic supplements have its major contributions, possibly because a particular strain is proven to improve a certain health condition and concentrating its presence in one supplement may help. If you don't have a particular serious health concern and would like to replenish good microbes to improve overall health, then maybe consuming fermented foods is a better option.

For the second circumstance, my thoughts are more in line with those of Sonnenburgs from *The Good Gut*. I prefer various food sources as there are varied sources of microbial profiles, whereas probiotic supplements usually come with only few specific types of microbes.

Considering there are "more than 10,000 microbial species occupying the human ecosystem", according to Human Microbiome Project, limiting yourself to a handful of species offered in a probiotic supplement will not be the most ideal way to cultivate a healthy and vibrant microbial community in the body.[41] Consuming different types of fermented vegetables and later moving onto varied types of fermented foods is likely a more efficient way to diversify microbial profiles and create a solid foundation for health.

But, there is really no right or wrong answer here. If you still want to try a probiotic supplement, go for it. As you decide what to buy, make sure that it: a) is produced by a reputable company; b) has clearly labeled the strains of bacteria included; and c) has a product shelf life listed. There is no evidence that any particular probiotic strain is more beneficial than the other; it is about taking your time to find out what is best for your unique microbiota, as long as you are sure the manufacturer is legit and trustworthy.

Despite my preference to consuming fermented foods to replenish microbes, some people with histamine intolerance might need to cut down on fermented foods and opt for probiotics. In the next question, I will talk more about this kind of intolerance in detail.

16. When should we NOT eat them?

Although there are amazing benefits you could enjoy by consuming fermented vegetables, it doesn't mean they are good for everybody, at least at this point. If you are currently suffering from histamine intolerance, it is actually a better idea to put fermented foods in general on hold.

Histamine is a chemical involved with immune response and proper digestion; it is a neurotransmitter that sends messages from the body to the brain. For those of you who don't know what that is or are unsure if you have this type of intolerance, it is generally caused by a defect in the body's histamine breakdown process, primarily associated with diamine oxidase (DAO) enzyme found in the intestine mucosa.[42] Typical symptoms include, but not limited to, itchiness of the skin, eyes, ears, and nose, hypotension (low blood pressure), nasal congestion, runny nose, seasonal allergies, fatigue, and chest pain.[43] Oftentimes, you may hear people say they experience itchiness in the skin after consuming fermented foods; which would mean histamine intolerance.

Foods high in histamine include, but are not limited to: alcohol, all kinds of seafood (such as shellfish or fin fish, fresh, frozen, smoked, or canned produce), fermented foods in general, cured meats like bacon, salami, pepperoni, dried fruits like dates, figs, and raisins, nuts, certain vegetables like eggplant, spinach, tomatoes, and avocados. [44]

Example low-histamine foods are vegetables other than those mentioned above, freshly cooked meat, gluten-free grains like rice and millet, and grain-like seeds such as quinoa, buckwheat, and amaranth, fresh herbs, and herbal teas.[45]

The good news is that histamine intolerance may not last forever; and it is possible that you can get out of this and start enjoying the benefits of fermented foods again. So, be sure to stay hopeful and optimistic.

As Schwenk pointed out during our talk, the base of histamine intolerance is lack of gut flora and a potential hint of adrenal stress. It is a warning sign from the body to ask you to start looking deep inside to fix what is not going well. So, if you think you have histamine intolerance, it is recommended you start fixing your gut first and work with a health professional to deal with this. You could try probiotic supplements, the effect of which is milder than that of fermented foods; or you could also consider juicing, which she believes is a great way to help heal histamine intolerance. Then, over time, you may be able to slowly reintroduce fermented foods into your diet again.

If, at this point, you still have other questions or concerns, you can send me an email at tracy@tracyhuang.me.

Now that you see fermentation can be a safe practice, let's officially move onto how to make your first batch of fermented vegetables. Before I take you to hands-on practices, we will do some orientation in the next chapter, so that you know what essential tools you need to prepare, and ideal environments for vegetable fermentation. I will also share some other nice-to-know getting-started tips as well.

Chapter Recap

- *It is not that acidic foods are bad and alkaline foods are good. Everything should be kept in balance. And you don't have to get too caught up by whether fermented vegetables are alkaline or acidic; rather it is the overall picture of having a plant-based diet that matters.*

- *People who suffer from acid reflux and small intestinal bacterial overgrowth (SIBO) can benefit from consuming fermented foods.*

- *Do not equal "being not used to the taste" to "not liking the taste" and be open-minded to unexplored flavors. The more you get used to the unique flavors of fermented foods, the more likely you will start to like them gradually. If fermented foods still taste* too sour for you, try mixing foods like sauerkraut with bland foods like unsalted quinoa.

- Despite a myriad of health benefits fermented foods have, you should always remember to consume them in moderation.
- At this point, you don't have to worry about nitrates and nitrites found in fermented foods, as long as you make sure that you make and consume them properly. Details can be found in Chapters 5, 6, and 7.
- Keeping ferments beneath the brine can effectively avoid molds from developing.
- It is recommended that you consume no more than 2,300 mg of sodium (about one teaspoon of salt) each day. Make salt-free fermented vegetables if you are already consuming too much sodium in your daily life and cut down your intake of processed foods.
- Naturally fermented foods are better than commercially fermented ones because the commercially made: a) may not have gone through a traditional fermentation process; b) may contain a lot of problematic additives like high fructose corn syrup, other forms of added sugar, sodium benzoate, polysorbate 80, and alum; c) can be pasteurized, which kills live cultures; d) and are usually placed on the shelf at room temperature. Instead, make your own fermented foods, which is: a) more cost effective; b) enables you to express creativity, be in control, and spread the love to whoever you are making them for.
- There are four key ways to ensure a safe experience in vegetable fermentation: mold control by keeping fermented submerged in brine, making sure pH reaches at least 4.6 before you consume ferments, enjoying fermented vegetables in moderation, and keeping sodium intake in check.
- Fermented vegetables are safe and even considered "particularly crucial" for kids and pregnant women. They are also gluten free, as there is no gluten in vegetables.

- *Both probiotic supplements and fermented vegetables have their own merits; it is hard to say which one is better. But I prefer getting microbes from various natural food sources because it ensures I get varied microbial profiles from different foods for overall health benefits, and because probiotic supplements usually contain only a limited amount of very specific strains.*

- *If you are currently suffering from histamine intolerance, it's recommended you heal gut problems first and opt for probiotic supplements. Then, slowly incorporate fermented foods later on.*

Exercises

- *Knowing that the recommended sodium intake is no more than one teaspoon a day. I think my current sodium consumption is _____ (too much/just about right).*

- *The list of processed foods I eat daily includes _____; to start taking control of my sodium intake, I plan to cut down _____.*

- *pH _____ is a key cut-off point to let me know when fermented vegetables are ready to serve.*

- *[If you plan to buy commercial fermented products] I will _____ to make sure that the product that goes into my basket is safe and free of toxic chemicals.*

- *[If you plan to try probiotic supplements] I will _____ to make sure that the supplement I choose will deliver the most benefits.*

- *If I have _____, I know I should heal my gut first and may start with a probiotic supplement; and I should be careful to consume fermented vegetables slowly.*

- *The four aspects to ensure consuming fermented vegetables is a safe experience are _____.*

Chapter 5:
Step Two –
Prepare Essential Items

"Do not be afraid. Do not allow yourself to be intimated. Remember that all fermentation processes predate the technology that has made it possible for them to be made more complicated."

– SANDOR KATZ
Wild Fermentation

Different books might teach you to use different tools before you ferment. While all advice is good; some tools and equipment just go slightly above the "necessary-for-beginners-to-get-started" line, meaning you might even skip those tools and equipment and still be able to make fermented veggies. Here, I am trying to cut down the number of tools you need to prepare so that you just keep the most essential. In the "Resources" section, you will find tools I use and recommend for your reference.

Essential Tools

A Kitchen Scale

You should consider whether you want to measure the vegetables by weight or by volume. If you measure by volume, the way you prepare different kinds of vegetables – chopped, minced, sliced, diced, or peeled by hand – can alter the results you are getting due to the differences in sizes and the amount of surface area. For example, Brussels sprouts take less space in a measuring cup if you cut them into small pieces. That's why measuring ingredients by weight is a better way to decide the amount of vegetables needed.

pH Test Papers/pH Test Meters

By now you have learned that it is important to make sure that the pH of ferments are kept at least 4.6 before you consume them. That's why keeping pH test papers or a pH test meter handy is necessary for newbies. Both papers and a test meter can give you a tangible and measurable way to monitor the fermentation progress, which is especially convenient for beginners.

pH papers usually come in two forms: two-inch strips and 15-foot rolls. Some measure a full range; this means its measurements range from pH 1 to pH 14. Others measure with some degree of accuracy by targeting a specific range. If you use papers, make sure that you find the ones that target with some degree of accuracy in the acidic range below 5. For example, you can find on Amazon some pH papers that measure the acidic ranges from pH 3.0 to pH 5.5.

Personally, I prefer using a test meter rather than test papers, because the former is more accurate: its reading gives you a number with two digits after the decimal point, whereas using pH papers only gives you one number after the decimal point. Having numbers with more digits after decimal points allows me to make better comparisons between different stages of vegetable fermentation and to evaluate the degree of acidity the fermented vegetables has dropped to.

Test papers can only show you colors; then you compare these colors shown on papers with a range of standard colors, each of which indicates a specific pH. With this method, you can only make an estimate of what the pH is. On the other hand, a well-calibrated test meter can give you a more accurate pH result.

Besides, as I let one drop of brine drip onto one end of a test strip, I found that pigments in ferments might sometimes influence the color of the result. For example, the color from red cabbage may increase the redness of the result; and the color from leafy greens may enhance blueish color. That's why I think using a test meter gives you a more effective way to measure.

To learn more about how to use a pH test meter and the pH paper, you may visit www.tracyhuang.me/ph-meter-and-paper.

A Thermometer

As you will learn later, temperature is one of the most crucial factors you should be in control of while you ferment. Knowing what temperature the fermentation process starts is key to producing a batch of flavorful fermented vegetables that please your taste buds. Chances are you may have a thermometer at home already, as the control panel of the heating and cooling system in your condo or house usually reflects the room temperature. In case you don't have one, you will have to buy a thermometer before you start making fermented vegetables.

Knives

Ideally, you will need to prepare two types of knives for cutting veggies: one 10 – to 12–inch knife for cutting large sized vegetables like cabbages, squashes, and cauliflowers, and the other one should be a small-sized chef's knife for chopping, slicing, dicing, and mincing smaller items like carrots, garlic, and ginger roots.

In fact, you may not even need to use a knife (you can pickle entire baby cucumbers or carrots or use your fingers to tear cabbage leaves into pieces). However, to help you enjoy the fun and variation of ways to prepare and consume fermented vegetables, keeping knives handy enriches your fermenting experience. What's more, you may have these two types of knives in your kitchen already; so, there is no need to make another trip to the supermarket. You may even only use one knife, as long as it performs nicely the two types of tasks mentioned above.

A Cutting Board

Knives and a cutting board go together. Like knives, a cutting board may not be needed as you make your first fermentation experiments. But, as you go further into the book, you will learn more ways to prepare vegetables other than peeling cabbage leaves with your fingers. You will also learn that different ways to prepare vegetables with knives and a cutting board can add to the joy and fun of making fermented veggies as well as changing the textures of these veggies as you taste them. That's why keeping a cutting board is necessary.

A Peeler

If you want to ferment root vegetables, a peeler comes in handy. You will need it to peel off the outer skin of vegetables such as carrots, daikons, potatoes, and beets which may be dusted by dirt.

Containers

There are many options that serve as containers to hold vegetables. Choices include water-seal crocks, onggi pots, glass jars, and food grade buckets. If all these words sound like Martian to you, you don't have to worry at all. It is likely that you are completely new or only have very limited knowledge of this topic. A lot of the containers mentioned here are for commercial use because of their large sizes.

As it is likely that you are making your first batch of fermented vegetables at home, I would recommend you simply get started with glass jars to make your first batch.

Mason jars are a good investment. They are relatively cheaper. What's more, as they are made of glass, you can easily observe what is happening inside. It is easier for you to check for "air pockets", the air stuck between ferments inside the bottle, and whether the ferments float to the top of the brine and need pressing down.

What I also like about glass jars is that they come in different sizes. The good news about this is you can start with small-sized jars to begin fermenting those vegetables which you are not sure if you would like. Some people worry: "I'm mostly afraid of how it would taste!" Isn't this an economic way to kick off your journey, so that you don't regret "wasting" the vegetables and so much of your time making a big batch that you end up not enjoying? It is after you gain confidence in the taste, flavor, and texture of a small batch when you start to go for more ambitious and adventurous fermentation projects by using larger-sized glass jars.

From Chapter 3, you have learned that chlorophyll-rich vegetables can develop strong flavors that might not appeal to everybody. After knowing this, I intentionally started my kale fermentation with a ½ pint glass jar to see if I liked that taste. As expected, the ferments generated a strong smell and flavors that I actually didn't like that much. As this may not be everyone's cup of tea, I, therefore, would not recommend kale fermentation within your first batch of experiments.

Materials Used in Containers

As a side note, this is something that comes in handy, should you choose to go beyond glass containers for your ferments in the future:

What the container is made of matters a lot, as certain materials can react with ferments in the process. As a general rule of thumb, avoid fermenting foods in metal or plastic bottles. Specifically speaking, make sure that you avoid metallic containers that are made of aluminum, copper, cast iron, low-grade stainless steel, chrome, or nickel. With respects to plastics, the more malleable it is, the more easily that the bottle is going to react during fermentation. So what are the safe options to use? Besides glass, non-reactive materials include stoneware, wood, and ceramics.

A Tamper/Pounder (or, Its Alternative)

Fermentation is all about creating an oxygen free environment. Using a tamper or pounder to compact the shredded vegetables will help press out oxygen in the jar; it also helps bruise the veggies to further release brine. Alternatively, to save a few bucks, you can explore around your place to dig out what you could use instead to achieve the same results. For example, the tamper from your Vitamix blender, if you have one, can do the work. If all else fails, you can always try this last resort – using the strength of your fingers to press down the shredded veggies, particularly when you are using small-sized glass jars.

Followers

Just because you pour brine over the shredded vegetables into the glass jar, it doesn't mean it creates a 100% oxygen-free environment – some shredded pieces might float to the top of the brine, especially at the initial stage when carbon dioxide is created in the fermentation process and gas is trying to get out, which makes the previously packed veggies less compacted. To prevent that from happening, followers are needed, which are used on the top of pressed ferments to keep them underneath the liquid in the bottle. Cabbage leaves can serve as ideal followers because of their rubbery texture, which aids the leaves in clinging onto the glass wall of the jar to help them "lock" themselves where they are.

There is a down side of using vegetable leaves – leaves shrink as the fermentation process proceeds and, therefore, may not be able to serve as a follower to keep ferments underneath the brine; they may even float to the top. The good news is that you can take out the shrunken followers and add in fresh new ones to keep ferments submerged; and, the fermentation process will not be influenced, as long as you make sure that the new followers added later on are clean.

When I was reading *Fermented Vegetables* by Kirsten and Christopher Shockey, I learned that Ziploc bags filled with water and sealed well can add weight on the top of ferments and, therefore, serve as followers as well. But, wait: aren't Ziploc bags made of plastic? And plastics can react with acids? Due to my concern about whether Ziploc bag materials would react to acids formed during vegetable fermentation, I conducted some research and below is what I have found:

Thankfully, using Ziploc bags can be safe. Ziploc bags are made of polypropylene (PP), known for high levels of chemical resistance.[1] Under lower temperature (about 67 °F /20 °C), PP demonstrates excellent resistance to acetic acid and lactic acid, two major acids formed during vegetable fermentation processes.[2]

To quote directly from the report posted by a Cornell University site: PP causes "no damage after 30 days of constant exposure (to both acids)". It is when both acids with higher concentrations (85% or higher) are exposed to higher temperatures (about 122 °F; or, 50 °C) that both may pose "little or no damage after 30 days of constant exposure."[3]

From an experiential perspective, Shockey told me in an interview that she liked using water-filled Ziploc bags as followers because, as she mentioned, "the bag can work so well." Besides filling the bags with water, you can also use brine or pebbles.

Coverings, Lids, and Airlocks

Adding a lid on the top of a jar helps prevent ferments being exposed to air and keeps out bugs or other foreign objects. Usually, you will find three types of lids for Mason jars: plastic ones, metal ones, and airlocks. Considering materials of lids (plastics and metal, specifically) can be reactive to acids formed during fermentation process, it'd be better if you can leave one to two inches of space between the top of the brine and the rim of the jar.

Of course, a covering is not always needed. You can fill a small glass jar with water and stuff it into a wide-mouth jar to press down ferments and keep them submerged. In this case, you may not need a lid. Besides, if you choose to use a Ziploc bag filled with water, brine, or pebbles as your follower, the bag may expose partially its body out of the jar (as the room inside the jar may not be able to fit the whole bag); in this case, you may not need a covering either. However, since this is likely to be your first batch of fermented vegetables, it's recommended that you use a smaller-sized jar which may not have a mouth wide enough to fit in one small glass jar. If you don't use a Ziploc bag as a follower, I still recommend you keep some lids or airlocks.

I have included a step-by-step guide on assembling an airlock. You can visit www.tracyhuang.me/airlock for the details.

A Deep Bowl

A deep bowl can come in handy when you need to mix several ingredients to blend different types of vegetables together nicely. It is also a good place to sprinkle salt on the top of vegetables and mix them up to help the salt penetrate the veggies and drive out the brine which you will use later on, if you choose to use salt.

A Pen and Sticky Notes

Personally, I found this particularly important and a must, especially when you try to make several batches of fermented vegetables at the same time. You will need to note down what you are fermenting, the current dates, and projected dates of completion. If you are feeling fancy, you can try to set up a Google document to keep track of this instead of using the good old fashioned way. Either way, the key is to keep track of your experiments. Take the example of when I made 16 bottles of fermented vegetables, I labeled each bottle with a number on a small piece of a sticky note, and set up a Google Excel Sheet to jot down current and completion-dates.

A Container to Hold the Packed Jar(s)

What may happen is that the brine may leak after the fermentation process starts, as gas produced may try to escape. Using a baking sheet or plate may do the trick. Then, observe it daily if possible (or at least for the first few days when fermentation becomes most active). If there is any leak, adjust the brine level, and if the container you use is metallic or plastic then rinse the lid as soon as you can to minimize chemical reactions.

A Journal (Optional, but Recommended)

Although it is labeled "optional" here, I still recommend you keep a journal of details on your learning process, because reflection makes you grow and helps you produce better fermented vegetables for yourself and your loved ones. Specifically, you can jot down your observations day by day, mistakes you have made, lessons you have learned, things you have done right, and inspirations that come up that can add to your next experiments.

This is also a place where you paste visuals of different stages your fermented vegetables have gone through. Starting out, I documented the whole process and observations along the way, which helped a lot with my learning. That's why I highly recommend this. You can find a template of how I documented my experiments by visiting www.tracyhuang.me/fvm-resources.

STEP TWO – PREPARE ESSENTIAL ITEMS

Nice-to-Have Tools

Besides the most essential tools, two that can be handy to speed up your food preparation process are a food processor and a garlic mincer. When you want to mix a lot of spices like red chili peppers, ginger, garlic, and jalapenos into your ferments, a food processor helps you complete that task fast.

It is also good to keep a kitchen cloth big enough to cover all bottles of ferments you've finished. As I was making bottles of fermented vegetables, the 4pm sunlight usually penetrated through the window and found itself directly resting on my bottles. I would then either transport the bottles a different spot; or, cover them with a big-enough kitchen cloth to keep them from direct sunlight to avoid temperature fluctuations.

By now you have learned what kinds of tools and goodies you need to keep before you officially start out your fermentation process, so let's now study the basic environments we need to set up for a successful fermentation experiment. To get started, I categorize all factors you need to consider into two big categories: external environments including temperature and sunlight control, and internal environments including oxygen control, timing, water quality, acidity level control, cleanliness, and saltiness of brine.

External Environments

Temperature
As stated earlier, temperature is the one of the most important crucial factors you should consider before you start fermenting. Shockey from *Fermented Vegetables* mentions that temperature control is important because the right temperature creates an ideal environment to slowly and properly develop acidity and flavors.[4]

Therefore, to set yourself up for success, you need to consider putting your ferments in the recommended temperature range. Different experts have slightly different opinions on what the ideal range should be. But, in general, it is between 50 °F (10 °C) and 75 °F (roughly 24 °C); a temperature between 50 °F and 65 °F (about 18 °C) is considered the most ideal. And, according to Caroline Barringer, Nutritional Therapy Practitioner and Vice President of the Nutritional Therapy Association, temperatures should not go beyond 80 °F (roughly 27 °C).[5]

If the temperature is too warm in your location, you can try to leverage the use of ice packs and a cooler to take control of the temperature.

Sunlight Control

It is important to keep the bottles of packed fermented veggies away from direct sunlight, because "(it) could cause light damage and temperature fluctuations in your crock and thus disrupt the bacteria's work," as mentioned in *Fermented Vegetables* by the Shockeys.[6]

Internal Environments

Oxygen Control

Now let's shift our attention to inside the bottle and take a look at how to expel oxygen out of the shredded vegetables, which is a fundamental step for vegetables to be successfully fermented. Always remember, if you are not keeping veggies submerged below the brine, you are not fermenting them properly. That's the key. The way you do it is by firmly pressing the prepared vegetables down to the bottom of the jar and by keeping them submerged in brine.

Water Quality

The type of water you use can also influence the quality of your ferments. Chlorine, which can easily be found in tap water these days, can kill off the microbes in the fermentation process. So, make sure that you take extra steps to remove chlorine before you ferment. To test the chlorine levels in water, you can buy chlorine measuring kits at your local pool supply shops. You can remove chlorine by using filters, boiling the water, or leaving water in an open pot for natural evaporation. Alternatively, you can just buy chlorine-free bottled drinking water.

Acidity Level Control

As mentioned earlier, pH 4.6 is your benchmark to decide whether ferments are safe to consume. Interestingly, Shockey also covers in her book *Fermented Vegetables* that the pH reaching below 4.6 is when the ferments are ready and taste good.[7] It seems that the pH reaching 4.6 is a defining moment to deliver both safety and flavors in fermented vegetables. Therefore, I recommend you keep pH test papers or a pH test meter handy to help you determine the right point in time to enjoy tasty and healthy fermented vegetables.

Timing

It is hard to say how long it takes before the ferments are ready to serve, as one type of vegetable fermentation can be different from the other. Instead of relying on the amount of time to decide when ferments are ready, I find that testing their pH levels is more straight-forward and easy to measure.

As long as the pH reaches 4.6 or lower, you are sure that the batch is safe to consume. In other words, the time for the bottle of ferments to reach 4.6 is the minimum amount of time you will need to wait for. Using this standard, I learned from my fermentation experiments that some ferments like fermented beet juice, sauerkraut, and kimchi could be consumed in as few as four days, as their pH dropped quickly. The pH of chlorophyll-rich greens like kale dropped slowly and took longer time to ferment; my guess is that greens are more alkaline in nature; and, therefore, it takes more time to reach an acidic level.

Cleanliness

Of course, just as you wash your pot before you cook anything, you need to clean the non-reactive container before you officially start the fermentation process. However, do you still remember what you learned earlier about how people's obsessive cleanliness in modern days are killing the good microbes? You don't need to be too fanatical about the cleaning process: for example, sterilization is not necessary for fermenting vegetables. Keeping the container clean simply by rinsing it should be a good start.

The Science of Salting

Despite the need to avoid excessive sodium intake, adding a certain amount of salt in your prepared ingredients actually has a myriad of benefits. To start, mixing salt with prepared veggies can help release juices from vegetable cells, which aids vegetables in creating their own brines.

Next, as mentioned above, salting helps increase the crunchiness of the vegetables. This is because salt can harden the pectin in the vegetable cells, which makes veggies crispy.

There are more benefits that follow: to maintain a certain degree of sodium level inside the jar helps save salt-tolerant bacteria like LAB and inhibits the growth of disease-causing bacteria and yeasts which will break down the sugars into alcohol instead of lactic acid. According to *Fermented Vegetables*, 0.8% of salt to veggie weight can prevent this kind of decomposition.[8]

Although it is primarily the lactic acid and the low pH environment that form the powerful preserving agent to inhibit pathogens, salt can also serve as a form of preservatives to help protect the ferments.

Besides that, the addition of salt can slow down the fermentation process, which means a higher quality of fermented vegetables to be produced.

Now that you understand why salting is important, we'll move onto how to implement the science of salting. Generally, the salt you need weighs about 1.5 – 3% of the weight of the veggies; for example, 1.5% of salt content is standard for kraut making; and 3% is ideal for making pickles and kimchi. If the salt percentage is too high (the warning line is 10%), it can inhibit fermentation process.[9]

I've created the following formula to help you quickly find out the amount of salt you need based on the weight of your vegetables and the salt percentage you choose to go for. It is:

vegetable weight in pounds × 454 g/pound × desired % × 0.2 teaspoon/g of grinded sea salt = desired teaspoon of salt you need

As an example, if your vegetables weigh one pound and you would like to go with 1.5% of salt content, you will need about 1.3 teaspoons of salt (1 pound x 454g/pound x 1.5% x 0.2 teaspoon/g).

After learning why salting is important and how to add salt in a scientific way, the next you need to know is how to select the right salt.

In generally, opt for mineral-rich salt. Himalayan pink salt is a great option. If possible, avoid using refined salt with added iodine, as it may inhibit fermentation and cause a discolored product.

As a side note, if you are seeking ingredients for salt-free brine, try kelps, seaweed, dill seeds, caraway, and celery stalks to introduce sodium.

Make Fermented Vegetables Your Own Way While Playing It Safe

Above I have covered the fixed rules you have to follow to set up ideal internal and external environments in-order to ferment vegetables successfully. However, fermentation wouldn't be fun if there wasn't room for creativity. That's why I've come up with the following summary to help you control different aspects of the fermentation process tailored to your own preferences. Yes, you can stay disciplined while being creative, as long as you operate within the parameter then you can stretch your creative muscles to try out different experiments at your own will.

Controlling Speed

The speed of fermentation can be controlled by manipulating the temperature and saltiness of the brine. In general, the higher the temperature there is, the faster the fermentation process occurs. Yet, remember that you can kill lacto bacteria if the temperature is too high. As mentioned earlier, an ideal range is between 50 °F (10 °C) and 75 °F (roughly 24 °C); and it is not recommend that the temperature go beyond 80 °F (roughly 27 °C).

By adjusting the amount of salt content in your brine, you can also control the speed of fermentation process. The more salt added into the brine, the slower the fermentation process. However, as mentioned before, it's not suggested to go beyond the recommended allowance of one teaspoon of salt per day. So, use your discretion as you adjust salt levels to adjust the speed.

Controlling Flavors

Besides adjusting the sourness, there are many other ways to please your taste buds. One way to adjust the saltiness of the brine and fermented vegetables is by adding in different amounts of salt. The general rule of adding salt is the same as that used for adjusting the speed of fermentation, as mentioned above.

Another way to enhance the flavor of the batch you are making is to mix them with different types of spices such as hot peppers, garlic, onions, scallions, shallots, leeks, caraway, oregano, turmeric, cumin, black peppers, coriander, fennel, mustard seeds, basil, chives, and parsley. Not only can these herbs and spices add dynamic new flavors, but they can act as mold inhibitors, as well.

Controlling Texture

There are four elements that can help you change the textures of the fermented vegetables:

First, timing, In general, the longer it takes, the more likely the fermented veggies get mushy. If you want to enjoy fermented cabbages while the leaves are still crunchy, you might not want to let your vegetables stay for too long.

The next factor is the amount of salt added. A proper amount of salt added into the shredded vegetables can increase the crunchiness. More of the benefits of salting will be introduced later.

Next, it is the way you prepare your vegetables. As you may have experienced already, shredded cabbages may taste differently in texture from cabbage leaves peeled by hand; and, similarly, thin slices of carrots chewed in your mouth feel different from diced carrots. Based on your preferences of texture, you can prepare the ingredients in a creative way, such as shredding, chopping, hand-peeling, pulsing, dicing, grating or mincing, and slicing. Or, you can even throw the whole piece – like a baby cucumber or a baby carrot – into the jar, if there is enough room.

The last trick is to add leaves of perennial shrubs, vines, and trees, because they contain tannins that can help the ferments remain crispness. Examples are black tea leaves, grape leaves, and oak leaves.

Other "Getting Started" Knowledge

Now that you have learned the most essential tools you need and you know a thing or two about the favorable environments for successful fermentation, there is still some basic knowledge you need to learn before you officially get started; and I've grouped everything you need to know in this part.

While you can use starter cultures for your fermentation experiments, it is not a must. You can ferment vegetables nicely even without the use of starter cultures. When you go out to buy ingredients, try to keep them as local and fresh as possible; make sure that you avoid produce with high pesticide levels; organic foods are ideal.

Besides shopping at local farmers' markets and grocery stores that supply natural foods, you can also buy organic vegetables by joining Community Supported Agriculture (CSA), a subscription-based program to get organic produce, know where your foods are coming from, and support local economy at the same time. If you have the space and bandwidth, why not grow your own foods? By doing so, you can also cultivate your microbial community by playing with dirt and immersing yourself in nature.

Furthermore, maintain basic hygiene by cleaning your work surfaces, tools, and hands with warm soapy water.[10] At last, rinse vegetables in cool water without using soap.

Is Using Starter Cultures Necessary?

Starter cultures can speed up the fermentation process by adding bacteria and sugar (a lot of the freeze-fried starter cultures contain sugar as well). This leads to the pH dropping quickly. As a result, the sudden drop in the pH can influence the stability of the acidity levels, according to Breidt. In other words, the pH can rise again even after it is well below 4.6.

While it is totally fine to use starter cultures for vegetable fermentation, you will have to make sure that the pH of the vegetables is stable. That's why checking the pH of the fermented vegetables is recommended.

I personally prefer using just salt and vegetables for a slower but steady fermentation process. As a matter of fact, it is not as "slow" as you think; fermented vegetables made with these two ingredients can be ready to serve in around three to five days, which you will learn more about in Chapter 12.

Success goes to those who are well prepared. By now, you understand the power of consuming fermented foods and healing your gut; you have removed your mental blocks and misunderstandings about fermentation, and learned all the necessary tools and conditions you need to know. You are well on your way to making your first batch of fermented vegetables successfully. In the next chapter, you will learn nine procedures that you always have to consider before picking the type of fermented vegetables to make. Of course, I will also cover details of making each type and personal lessons learned during my vegetable fermentation experiments.

Chapter Recap

- *Essential tools to keep for successful vegetable fermentation:*
    - *a kitchen scale*
    - *pH test papers or a pH test meter (the latter is preferred)*
    - *a thermometer (or, a way to know the temperature of where you are making fermented vegetables)*
    - *one knife or two to be able to cut cabbages and thick-skinned squash as well as chop, slice, dice, and mince smaller ingredients*
    - *a cutting board*
    - *a peeler*
    - *glass jars of different sizes*
    - *a tamper or its alternatives*
    - *extra cabbage leaves or Ziploc bags filled with water, brine, or pebbles as followers, lids or airlocks, a deep bowl, some tape and markers*
    - *a container to hold the packed jars I case of brine leak*
    - *a journal (recommended) to keep track of your fermentation progress and jot down observations and lessons learned*

- *Nice-to-have tools: a food processor, a garlic mincer, and a kitchen cloth big enough to cover all bottles of ferments you make.*

- *External environments to consider: recommended temperatures range from 50 °F (10 °C) to 75 °F (24 °C) and should never go beyond 80 °F (27 °C); stay away from direct sunlight to avoid light damage and temperature fluctuations.*

- *Internal environments to consider: always remember to keep ferments submerged in the brine; avoid chlorine in water; make sure that the pH of ferments reaches at least 4.6; the time it needs for ferments to reach pH 4.6 is the minimum amount of time you will need to wait for; clean the glass jars without sterilization; it's suggested that you start with 1.5% salt content for making krauts and 3% for making pickles and kimchi.*

- *There are many ways you can be creative about vegetable fermentation: you can control temperatures and saltiness to control the speed of fermentation; the longer you ferment, the more sour your ferments will get; you can add in different herbs and spices to enhance the flavor; the texture of ferments can be influenced by the amount of time you ferment, whether or not you add salt into the brine, how the vegetables are prepared, and whether you have added leaves from perennial shrubs, vines, and trees.*

- *Proper salting can help release juices to prepare the brine, produce crunchiness, save salt-tolerant lactic acid bacteria, serve as preservatives, and slow down the fermentation process to enhance quality.*

- *How to salt: it's ideal to add salt at 1.5% to 3% of the weight of the vegetables and do not exceed 10%.*

- *Salting formula: vegetable weight in pounds x 450g/pound x desired % x 0.18 teaspoon /gram of salt = desired teaspoon of salt you need*

- *What salt to use: mineral-rich salt is recommended; avoid refined salt with added iodine.*
- *Other nitty gritty details: using starter cultures is not a must; always opt for local, fresh, organic vegetables, if possible; clean work surfaces, tools, and hands with warm soapy water; rinse veggies in cool water without soap.*

Exercises

- *After browsing all essential tools, what I don't have in my kitchen is/are _____; I plan to go to _____ to buy them.*
- *The minimum acidic level required for a ready-to-go batch of fermented vegetables is a pH of _____; ideally, the pH can reach _____ or below.*
- *If I plan to make one pound of cabbage with 2% salt, then I need to add _____ teaspoons of salt.*

Checklists and a Template

- *Visit www.tracyhuang.me/fvm-resources to find out a template of how I documented my fermenting experience, along with these checklists: "Essential and Nice-to-Have Tools", "Factors Crucial for Vegetable Fermentation Success", and "How to Make Fermented Vegetables in Your Own Way While Playing It Safe".*

Chapter 6:
Step Three – Make It Happen!

"Fermentation is easy and exciting. Anyone can do it. Microorganisms are flexible and adaptable. Certainly there is considerable nuance to be learned about any of the fermentation processes, and if you stick with them, they will teach you. But the basic processes are simple and straightforward. You can do it yourself."

– SANDOR KATZ
Wild Fermentation

In fact, fermenting vegetables is not overwhelming at all. According to Katz, any kind of vegetable fermentation method follows just one golden rule: "by keeping vegetables submerged under liquid, you create a selective environment where molds and other oxygen-dependent organisms cannot grow, thereby encouraging acidifying bacteria.[1] Here is another quote from *Fermented Vegetables* that is a pithy little saying to keep in mind: "submerge in brine and all will be fine."[2]

After my studies and experiments, I have learned that whatever ferments you want to make, the overarching process is basically the same. That is, you should always consider these nine procedures whatever ferments you decide to make: first, choose vegetables (ideally, they are fresh, local, and organic); next, prepare vegetables; third, mix salt into prepared veggies, if necessary; next, pack the mixture into the jar; fifth, add the brine if needed; sixth, add followers on the top of all ingredients to keep them under the brine; seventh, add a lid on top; eighth, place the jar into a bowl to hold liquid in case of the brine leaking out, which is followed by the last step: to monitor the whole process, especially during the first few days when fermentation is most active and to make adjustments accordingly.

Yes, you think through these nine steps over and over again.

It might seem daunting at first. The good news is, if you are able to see this pattern, you will find that fermenting vegetables is not overwhelming at all, as it is just a flow of repeated steps you need to follow for whichever type of fermented vegetables you want to make.

On the other hand, this may appear "boring" to some people. However, there are small tweaks for each type of ferment – such as varying the type of vegetables you choose, changing the ways you prepare them, adding different amounts of salt for different types of ferments, adjusting salt content in the brine, using different kinds of, and incorporating your own creative ideas, into the process, etc – that makes the seemingly monotonous procedure fun, enjoyable, and rewarding.

Based on what we've learned above, I've collected 16 established recipes – combined with minor tweaks – just enough to cover all types of fermented vegetables that I'm about to share below. Recipes can be found in Chapter 12. I've incorporated my personal learnings from my trials and errors as I describe to you how to make different ferments.

Before we begin, note that if you use a Ziploc bag, you can consider skipping the step of adding a lid to the jar. That's because a Ziploc bag can not only serve as a follower to keep ferments submerged, but also keep the air from getting into the jar as well. You can refer to the instructions on how to make bases and pastes and how to make fermented vegetable leaves to get a step-by-step guide on using a Ziploc bag as a follower.

Making Krauts

The basic guideline for kraut making is letting vegetables produce their own brine *without* adding more externally. Brine making is often helped by the addition of salt to bruise leaves and other ingredients and produce juice. And, do not only limit yourself to just sauerkrauts.

Besides krauts made from cabbages, commonly known as sauerkraut, other types of krauts include those made with other vegetable combinations other than cabbages, with just whole leaves, and with a small portion of vegetables mixed with a large proportion of water (this way, you are making fermented juice in the end). I will discuss each type in detail below.

Sauerkraut

1. Key Principle

The key to making sauerkraut is to use red or white cabbages as the main ingredient and follow basic principles of kraut making outlined above.

2. How to Make Sauerkraut

When choosing the cabbage, make sure that it is a firm head and has crisp and shiny leaves. Alex from *Real Food Fermentation* teaches that you should remove the outer leaf if the cabbage is not organic.[3]

Then, you cut the cores out; third, you can either shred cabbage leaves into strips or hand peel them into smaller pieces. Once you are done, place them into a bowl.

The next step is to decide how much salt you want to add. Once again, this is based on your understanding that: A) everyone should keep their sodium intake in check; B) about 1.5% of salt per pound of vegetables is ideal for kraut making; and C) the salt you are about to add should not weigh more than 10% of the weight of the prepared vegetables.

Mix the salt into the prepared vegetables and massage with your hands and fingers for a few minutes, set the bowl aside, and wait for about 30 minutes until the brine is formed. Alternatively, you can continue to massage the prepared cabbage until the brine is produced. It took me about eight minutes to massage one pound of shredded cabbage (mixed with two teaspoons of sea salt) until the brine was formed.

Next, pack the prepared vegetables into the jar as tight as possible by either using your fingers or a tamper to press downwards. The key is to leave out the air and create an oxygen-free environment. As you firmly press against the vegetables, you will see that the brine starts to appear while the veggies are submerged underneath.

Carbon dioxide will be generated as soon as the fermentation process begins. As a result, you may start seeing bubbles hidden in the veggies as early as the next day.

The vegetable leaves may expand and be exposed to the air, preventing you from a successful fermentation. That's why it is important to add several layers of followers on the top of the packed veggies to keep them submerged. A very simple method is to save a few pieces of cabbage leaves right before you start chopping up the whole cabbage or hand peeling the cabbage leaves.

Firmly press the followers into the mouth of the jar against the top of the shredded veggies and make sure that they are all submerged underneath the brine. The rubbery texture of the followers help themselves cling to the jar wall, so that they "stick" to the wall and will not be easily pushed upwards as carbon dioxide is generated and escapes from the veggies underneath.

Leave one to two inches between the top of the brine and the lid to avoid the brine leak and chemical reactions between the brine and the lid.

After that, you add a lid on the top. Mason jars usually come with two types of lids: one is made of plastics; and the other, metal. Since these two types of materials can be reactive to acid generated from fermentation, you have to be very careful that the brine is kept away from the lid by a few inches apart. Not knowing this earlier, I once carelessly let the brine stay in constant contact with the lid through the fermentation process. The consequence was that my metal lid became slightly eroded.

Here is one more important thing to remember: you don't want to seal the bottle so tight that no air can escape. At the initial fermentation phase, carbon dioxide will be generated; if the bottle is sealed too tight and carbon dioxide is trapped from within, this will generate pressure inside the bottle, which can make it hard to open the jar later on or cause bubbles to burst out forcefully as you open the jar. As a result, "let it naturally just stop turning", as Caroline Barringer tells Dr. Mercola in an interview, without turning it further by force to make the sealing more tight.[4]

Next, if you are making multiple jars at the same time, I recommend you add a sticky note or label somewhere on the jar for differentiation and process tracking.

When all the above is done make sure to prepare a bowl big enough to hold the jar and place the jar into the bowl in case any brine leaks out.

You can visit www.tracyhuang.me/sauerkraut for the visual guide.

3. Lessons Learned

I made two versions of sauerkraut – one was salted; and the other, not salted – to note the differences and found that the salted one enhanced the flavor and crispness of the ferments, as expected. With slightly more than 1.5% salt, the end product smelt more pleasant than the salt-free sauerkraut and tasted better, in my opinion.

In the meantime, be mentally prepared that making krauts with a red cabbage may require more labor. When using red cabbage instead of white cabbage, I found that it took much longer time to produce the brine. For an eight-ounce shredded red cabbage with 1.5 teaspoons of sea salt, I needed to massage it for 10 minutes or more before seeing any brine, whereas it only took eight minutes for one pound of shredded white cabbage with two teaspoons of salt to form enough brine to submerge it.

I later learned from Shockey that: when there isn't enough brine formed, there are tricks you can do to add more brine into the ferments to keep them submerged. Had I known this as I was making the red cabbage kraut, I would have considered doing what she suggests in her book: 1) add more salt and keep on pressing down firmly; 2) add vegetables that can give off more liquid, like grated or shredded turnips; and 3) add citrus like lemon juice, lime juice, grapefruit juice, or leftovers from other fermented juices. And, she does *not* recommend directly adding salted water, which can lead to mushy krauts.

Kraut Variations

1. The Key Principle

After knowing how to make sauerkraut, you are now well on your way to conquer the making of kraut variations. The main difference between sauerkraut and other kraut variations is that the latter combines a variety of vegetables besides cabbage as well as various kinds of herbs as you like. Then, you simply change the sizes and shapes of the vegetables and decide the combination of spices and herbs. To make kraut variations, you will still use the brine produced by the vegetables themselves.

2. How to Make Kraut Variations

Think beyond cabbages. You may go for all kinds of other vegetables like carrots, red radishes, cauliflowers, or leeks, just to name a few. I once fermented an eggplant, basil leaves, a carrot, and three garlic cloves by mixing them together in a jar and really enjoyed the mixed flavor.

Be creative about the ways you prepare them, as long as they can fit into the jar, such as chopping, slicing, shredding, and slicing, followed by mixing all prepared ingredients together.

The method of making kraut variations is the same as the way to make sauerkraut: after vegetables are prepared, you will add salt according to the salting rule discussed earlier; you will then squeeze the salted vegetables with hands and fingers or let the mixture sit for about 30 minutes to form the brine.

Depending on the type of vegetables you use, you may need to massage for different amounts of time until enough brine is produced. For example, it took me quite some time for shredded red cabbage leaves to form just a little brine; it only took about three to four minutes for six-ounce chopped eggplant with one teaspoon of sea salt to produce the brine; it also took about three to four minutes for half a piece of a sweet onion with one teaspoon of sea salt to form the brine.

Next, you stuff the jar with colorful veggies of your own choice, topped with a follower which can be one or two small pieces of cabbage leaves. Add a lid without sealing the jar too tight and make sure that you leave one to two inches of space between the top of the brine and the lid.

If necessary, add a sticky note or label with dates on it on a visible spot to differentiate the bottles of ferments you are making. At last, place a bowl at the bottom to hold the brine in case it leaks out.

For the illustrations on how to make kraut variations, visit www.tracyhuang.me/kraut-variations.

3. Lessons Learned

As you decide what vegetables to use, also stay mindful about how chlorophyll-rich leaves like spinach and kale may generate a strong smell and flavor that not everyone will like. If you are feeling adventurous and would like to give it a try, I recommend you use a small jar (e.g. a half-pint jar) to begin your experiment.

Kraut Variations – Pastes and Bases

1. The Key Principle

Once again, the fundamental principle of making all kinds of krauts is same: vegetables with added salt will create the brine of their own and begin the fermentation process by being submerged into their own brine. The main difference here is that these krauts are pulsed into small pieces and may taste saltier than other types of krauts, because they are usually used as condiments or seasonings. Because of this, high salt content (around 3%) is recommended here.

2. How to Make Pastes and Bases

You first start off by having a variety of vegetables of your choice ready. Pulse them or chop them into small pieces and place them into a bowl. Decide the amount of salt you want to add and mix it with the prepared vegetables. Slightly massage the mixture, so that juices will come out immediately after.

The following steps remain the same: packing salted vegetables tight into the jar, placing followers on the top, and making sure that ferments are submerged in the brine and that there is one to two inches of space between the top of the brine and the rim of the jar.

Next, if you choose to use a lid, cover the jar with the lid without sealing the jar too tight to give room for carbon dioxide to escape. Add a label on the bottle to record basic information about the ferments. Lastly, place a bowl at the bottom to hold the brine that may leak out of the jar.

You may visit www.tracyhuang.me/pastes-and-bases for the step-by-step visual guide on making your first batch of fermented pastes or bases.

What If I Want to Use a Ziploc Bag?

You can use a Ziploc bag as another layer of followers on the top of the ferments by following these steps:
1. Open the plastic bag, tuck it into the jar, and let it sit on the top of the ferments.
2. Extend your fingers into the bottom of the Ziploc bag and firmly press the bottom against the top of the ferments to squeeze out the air pockets and keep the ferments submerged.
3. Withdraw your fingers, fill the bag with water, and seal the bag.
4. (Optional) Add a piece of cloth on the top to keep the ferments away from potential bugs or flies.

There is no need to add a lid on the top of the jar, because the Ziploc bag extends its body out of the jar. As long as you make sure that the ferments are well submerged in the brine, the fermentation process will proceed.

3. Lessons Learned

Depending on the size and the type of vegetable ingredients you choose, sometimes it may still help if you can massage the prepared ingredients to help the brine to form. In my experiment, to make fermented leek paste, I prepared 12 ounces of chopped leaks with half a teaspoon of salt and massaged the ingredients for eight minutes until I saw the brine appear.

I also found that the brine produced may not be able to help ferments submerge. If that's the case, you may try out one of the three tricks mentioned above to introduce more brine.

Kraut Variations – Whole Leaf Ferments

1. The Key Principle

As you can tell from the name of this type of ferment, whole leaves are the key ingredient. And you still apply the basic principle of kraut making.

2. How to Make Whole Leaf Ferments

After removing the leaves from the stems, you add salt and toss gently. Leaves will wilt and the brine starts to form as soon as you massage them slightly.

Then the rest of the process stays the same: you pack the leaves tightly enough into the jar and have enough juice to keep them submerged under the brine; add followers, while leaving space between the top of the brine and the rim of the jar.

If you use cabbage leaves as the followers, you then need to seal the jar with a lid, label the bottle with a date and other notes you want to jot down; and, lastly, place the jar into a container in case the juice leaks over the next few days.

You may check out www.tracyhuang.me/whole-leaf-ferments for the step-by-step visual instructions.

3. Lessons Learned

As green leafy vegetables may leave a strong smell and flavor that not everybody would like, it is recommended you start with a small jar to test if this is what you like. In addition, my experiment tells me that massaging can speed up the osmosis process: it only took one minute for six ounces of baby kale or basil leaves with half a teaspoon sea salt to shrink dramatically. And, be prepared to get surprised and see how a large bowl of baby kale or basil leaves can actually fit into a half-pint jar!

Kraut Variations – Fermented Veggie Juice

1. The Key Principle

Since this is a form of a drink, we expect to have a lot of liquid in the final production. And this is the kind of kraut making where adding water from the outside is a must. Fermented vegetable juices are easy to make once you know how to make sauerkraut and kraut variations with different kinds of vegetables. Basically, steps are very similar, other than the fact that you will need to add a lot of filtered water.

2. How to Make Fermented Veggie Juice

Like what you would need to do to begin making other types of krauts, you start out by gathering whatever raw ingredients you want to use; peel off the outer skin if necessary, and chop the ingredients into small chunks before throwing them into the bottom of a jar.

Next, sprinkle a few pinches of salt on the top of the chopped ingredients, followed by adding filtered water all the way to the top with one to two inches between the top of the brine and the rim of the jar.

Then, you seal the bottle with a lid, label the bottle with necessary notes, and place it inside a container.

As a reference, you may visit www.tracyhuang.me/fermented-juice to check out the step-by-step visual guide on how to make the fermented beet juice.

3. Lessons Learned

You will see that you do not need followers in this type of kraut making, as chunks of chopped vegetables can generate some weight and help them stay at the bottom without floating atop. As you prepare your ingredients, be aware that you don't cut ingredients into thin slices which can cause pieces to float to the surface.

Making fermented veggie juice is probably the easiest and least time-consuming kraut making method. Therefore, I highly recommend you consider this as one of your early experiments and to start with beets.

For my fermented beet juice experiment, the weight of the beet chunks naturally kept them at the bottom. So, there was no need for followers. High sugar content in beets allowed them to be fermented easily and quickly. And, pH dropped to 4.16 by Day 4. In other words, fermented beet juice was ready to serve after four days.

Making Pickles

The Art of Pickling

1. The Key Principle

Unlike kraut making, pickling requires adding the brine from the outside. In other words, you will have to make your own brine with higher salt content at about 3%. That is, you add one and a half to two teaspoons of salt per cup of water.

To produce better tasting ferments, you can also try using special leaves from perennial shrubs, vines, or trees as the followers, such as horseradish, grape leaves, or oak leaves. These special leaves not only keep vegetables under the brine but also help make vegetables stay crunchy, because they contain tannins that encourage crispness. Besides, you can also consider tannins-rich black tea leaves to achieve the same texture.

Cucumbers are not the only ingredient that can be used for pickling; other types of vegetables can be the ingredients, too, as long as you follow the basic rules.

2. How to Make Pickles

You will first need to gather ingredients such as vegetables, herbs, and spices.

Then, prepare them in desired sizes. In fact, you could place the whole pieces of vegetables, such as the entire baby cucumbers and baby carrots, into the jar if there is enough room. If you use cucumbers to make pickles, make sure that the blossom ends are cut off, as they contain an enzyme that will soften the pickles.

Next, pack the prepared ingredients into the jar as tight as you can, leaving as little air inside as possible. I used black tea leaves from a teabag as my source of tannins; and I placed them in the very bottom of the jar before I packed in everything else, as adding these leaves last may cause them to remain on the brine surface.

Once this is done, pour in the salted brine with about 3% salt content (around two teaspoons of salt per cup of water).

After you make sure that the ingredients are well submerged in the brine, you can add in special leaves of your choice as the followers. In my experiment, I did not add any followers on the top of the cucumbers. That's because if you pack enough cucumbers tightly together, this should be enough to keep them from floating to the surface of the brine.

You may see a little bit of tea leaves or bits of herbs float to the surface. In this case, you may scrape them out or just leave them on the surface. I personally found that the small amount should not influence the fermentation process.

Make sure that there is some space between the top of the brine and the rim of the jar, followed by adding a lid to seal the jar. Do not forget to label the bottle with a date and other notes, if necessary. At last, place the jar into a bowl and let it ferment.

You may check out www.tracyhuang.me/pickles for the visual instructions.

3. Lessons Learned

I made one bottle of cucumber pickles with these tannin-rich black tea leaves and made another bottle without them to test the difference in tastes. The result was that the one with black tea leaves indeed tasted slightly crunchier.

Pickles are not just cucumber pickles. I encourage you to try out pickles made from other vegetables other than cucumbers to enjoy the variety. I recommend you give bok choy a chance. From my experiments, these leafy veggies made very tasty ferments that were crunchy with pleasant smells.

Making Kimchi

The book *Fermented Vegetables* describes kimchi as "any vegetable pickled in the Korean style of lactic-acid fermentation". Traditional ingredients include: Chinese or napa cabbage, radish, garlic, scallions, ginger, and red chili peppers. As you can see from this definition, any type of vegetables, as long as you prepare them with the above approach, can become a form of kimchi. It is not limited to Chinese cabbage – what you commonly see in restaurants.

Different type of kimchi making

The making of kimchi lies between kraut making and pickle making. You can either follow the kraut making method or pickling method to make kimchi. A third way to make kimchi is the pickling method with an additional presoaking procedure. As you may tell from the name, there is an extra step to presoak ingredients before you officially start out to make kimchi. While you can refer to kraut making and pickling principles to perform the first two methods, we will dive into details of making kimchi with this third method.

1. The Key Principle

Like how you make pickles, you must prepare your own brine to make kimchi. As a reference, to make the kimchi brine, you would need about three teaspoons of salt per cup of water. You will also need to presoak the vegetables for 6-8 hours before the official fermentation. Just as you pickle with tannin-rich leaves from perennial shrubs, vines, and trees, the same types of leaves are recommended as followers to achieve the same results: to keep veggies under the brine and to keep them crispy.

2. How to Make Kimchi

In my experiment, I used the traditional ingredients – that is, Chinese napa cabbage and all kinds of spices – as an example to walk you through the whole kimchi-making process. You can also visit www.tracyhuang.me/kimchis for the illustrations.

To get started, you will need to have all the necessary ingredients in place including Chinese or napa cabbage, garlic, ginger, and red chili peppers. Then, you soak the vegetables in the brine; and set them aside at a room temperature for 6-8 hours. Next, drain the brined veggies briefly.

Afterwards, peel off two leaves and set them aside as the followers; chop the cabbage into one-inch pieces and put them in a large bowl and; chop or mince all herbs and spices into small pieces, and mix everything together and massage thoroughly until you see the brine is formed.

When this is done, pack the mixed veggies into the jar tightly. In my experiment, because I used black tea leaves as my source of tannins to enhance the crunchiness in kimchi, I added the leaves into the bottom of the jar first *before* I packed everything else; press down the vegetables as hard as you can to keep them submerged; add the followers on the top and, if necessary, add more brine to make sure that the followers are submerged as well.

Leave one to two inches between the top of the brine and the rim of the jar; loosely cover the jar with a lid to allow carbon dioxide to escape; stick notes to the bottle, if necessary; at last, place the jar into a bowl which is used to hold the brine that may come out during the active fermentation process.

3. Lessons Learned

I designed two experiments – kimchi made from the presoaking method and kimchi made from the kraut making method – to try to find out the difference. Interestingly, I found that the one made from the presoaking method tasted more sour and spicier than the one with kraut making method. It seems that presoaking can allow spices and juices to more easily penetrate into cabbage leaves, which enhances flavors of kimchi.

Meanwhile, just as you should go beyond cucumbers when you try to make pickles, you should go beyond Chinese cabbage as you make kimchi. So, actively seek out different ingredients that can make great kimchi with the pre-soaking method, kraut making method, and pickling method.

Can I Add Vinegar?

Another common question is whether or not vinegar could be added. The answer is yes! During my interview with Breidt, he mentioned that a little bit of vinegar lowers the pH, which helps create a favorable environment for fermentation; at the same time, a small amount of vinegar serves as a "buffer" to prevent the pH from dropping too quickly to allow a slow and steady fermentation. And, of course, you are also adding more flavors to the veggies, too.

What to Expect Next

The fermentation process begins as soon as all vegetables are submerged under the liquid. It is likely that you may be able to see bubbles start to form attached on the surface of the ingredients on the following day.

If you hear bloops coming from the lid, don't panic; that is carbon dioxide trying to get out to the jar. This is especially so when there is higher sugar content in the batch of vegetables (such as beets) and warmer temperatures. You may clearly hear them, especially during night time when it is quiet. If you make multiple bottles of ferments at the same time, you may even discover that different types of ferments may generate different volumes of sizzling. For example, radish, white cabbage, napa cabbage, beets, and cucumbers appear to make more noises compared with leeks, spinach, or kale.

It is important that you observe the progress frequently, especially for the first few days, when fermentation is the most active. At this stage, because of bubbling and gases trying to escape, vegetables may become loosely packed and they may go above to top of the brine. That's why you need to check in from time to time and, if that happens, suppress the veggies under the brine by pressing down on the followers.

You may see "air-pockets" within ferments – the small bubbles that get stuck among the ferments, which you can easily see through the wall of the glass jar. Having bubbles like these can influence the taste of final ferments. However, you don't have to open the jar to adjust this every time you see them. It's more important to minimize the introduction of the outside air and to prevent the disruption of the lactic bacteria work inside the jar. So instead, just squeeze out the bubbles, as many as you can, only when you *need to* open the jar (for example, when ferments are not fully submerged in the brine).[5]

You may see foams on top; but they are harmless. Should you see a small amount of mold on the top of the ferments, simply scoop it away and continue to ferment. Based on my experiments with 16 bottles, I didn't see any mold developed, although there were times when ferments and followers floated to the top of the brine. This shows to you that fermentation is not necessarily linked with mold growth.

Besides, do not get concerned if you see green vegetables like whole leaves, cucumbers, and asparagus turn a deep green wilted color which is muted and dull, as this is a natural progression that you will observe.

As you monitor the changes over time and make adjustments, here are a few things to keep in mind:

You should always remember to minimize the number of times you open the jar as you want the fermentation process to be disturbed as little as possible, especially at the early stage. Opening the jar can introduce oxygen and other bacteria that you don't want; the fermentation process could then be influenced, which might lower the quality of the fermented vegetables in the end.

You may open the jar more often when the pH of the ferments reaches the proper level discussed before (4.6 or lower) or, preferably, after you put them into the fridge which slows down the fermentation process immensely.

At the end of the day, you should use your own judgment to balance between not opening the jar too often and the need to open the jar to make adjustments. And always remember that when ferments float to the top and are exposed to air during fermentation, you will *have to* open the jar to keep the ferments under the brine.

When Do You Know It Is Done?

A large part of it is based on your personal preference. If you would like it to taste more sour, then give it a few more days and taste it daily until you reach the level of sourness you enjoy; if you like the texture of crunchiness, then take the batch out as soon as the pH reaches the recommended levels (pH 4.6 or below).

Besides, let your own senses make the decision for you by observing, smelling, and tasting the vegetables from time to time. Your observations can sometimes tell you the quality of tastes: mushy-looking sauerkraut often suggests less pleasant flavors and textures. If you open a bottle of ferments and they smell good, chances are they likely taste good, too.

You would have to lift the followers on top to taste samples to decide whether it is ready. If you feel that it is not yet ready, simply rinse the followers, put them back in place, and continue monitoring the progress. The purpose of rinsing the followers is to wash out impurities that may form during fermentation and to keep foreign substances outside the jar – like dirt in the air or bits of foods or foreign debris left on the kitchen counter that your followers may accidentally be in contact with. This practice can make ferments taste better in the end, too.

What Are the Creative Possibilities?

As long as you make sure that you follow the fixed rules, you can always develop your own creativity by tweaking multiple variables, most of which you have learned in the last chapter already:

By controlling the length of time, you get different levels of acidity; by adjusting the temperature within the range between 50 °F (10 °C) and 75 °F (24 °C), you can control the speed of fermentation, by experimenting with different sodium levels, you can control not only the speed of fermentation, but also the textures and flavors of the ferments. Further, always keep perennial plant leaves in mind and add them to different types of ferments to enhance crispness.

If you feel really adventurous, you can mix in different herbs and spices while making kraut, pickles and kimchi, too. For krauts, you can try the sodium-containing brine like celery juice.

What's more, I encourage you to try out different combinations of vegetables. You can choose to combine vegetables that share the same texture – like carrots and white radishes. You can also try mixing vegetables of different textures, so that different ferments from the same jar may compensate one another as soft ferments may produce more juice (like peppers), while crispy ones can enhance crunchiness for each bite. Besides, the addition of garlic and basil can often produce pleasant smells and flavors in ferments. You may also want to give them a try.

As you see, there is much room for creativity. Everyone should take advantage of all the suggestions mentioned above.

When a bottle of ferments is ready to serve, the next step is, of course, to enjoy it. However, many people have expressed their negative experience regarding consuming fermented foods. That's usually because they didn't know how to enjoy them properly. How should you enjoy fermented vegetables while minimizing the side effects caused by improper ways of eating them?

To answer this question, I have come up with a list of 12 areas you should be aware of to make sure that you reap the benefits of introducing fermented vegetables to your gut in a safe and proper manner, which will be discussed next.

Chapter Recap

- *The underlying principle for vegetable fermentation is to keep vegetables submerged in the brine.*
- *There is one similar pattern for every kind of vegetable fermentations that you need to consider, which includes: 1) choosing vegetables, 2) preparing vegetable, 3) salting, 4) packing, 5) adding the brine, 6) adding followers, 7) adding a lid on top, 8) placing the jar into a bowl, 9) monitoring and adjusting (if needed).*
- *The key principle for kraut making is that you let vegetables form their own brine with about 1.5% salt.*
- *The key principle for pickling is that: you must create your own brine externally with about 3% salt; which means when you prepare the brine, it's about 1.5 to 2 teaspoons of salt cup of water. It's recommended you add perennial tree leaves to help these ferments taste crispy.*
- *Traditional ingredients of kimchi include Chinese or Napa cabbage, radish, garlic, scallions, ginger, and red chili peppers. There are three different ways of making kimchi: the kraut making method, the pickling method, and the presoaking method. The key principle for making kimchi with the presoaking method is that: you must create your own brine externally, made with about three teaspoons of salt per cup of water. Soak Chinese or Napa cabbage for six to eight hours before officially making kimchi; it's recommended you add perennial tree leaves as well.*
- *What to expect after packing vegetables into the jar: you may see bubbles attached to the surface of the ingredients and hear sounds as carbon dioxide is trying to escape; vegetables may float*

to the top of the brine, which then requires you to push them back underneath the brine by readjusting the followers; when you see "air pockets" in ferments, you may want press down the ferments to squeeze out the bubbles at the same time as you are opening the jar for other tasks. It's okay if you see foams, as they are harmless; if you see molds in small amounts, just scape them away. Finally, green ingredients like whole leaves, cucumbers and asparagus will become deep green wilted color.

- *Minimize the number of times you open the jar, especially at the early stage.*
- *Fermented vegetables are ready to serve when their tastes and flavors meet your personal standard.*
- *Be creative about possibilities of making different kinds of fermented vegetables.*

Exercises

- *The first vegetable fermentation method I would like to try is _____.*
- *The key principles for performing this type of fermentation successfully are _____.*

Checklists

- *Visit www.tracyhuang.me/fvm-resources for these checklists: "What to Expect after the Fermentation Process Begins", "When to Know Fermented Vegetables Are Done", and "Creative Suggestions to Diversity Your Vegetable Fermentation Experience".*

Chapter 7: Step Four – Eat Fermented Vegetables the Right Way

"Lacto fermented vegetables increase in flavor with time – according to the experts, sauerkraut needs at least six months to fully mature. But they also can be eaten immediately after the initial fermentation at room temperature."

– SALLY FALLON & MARY ENIG
Nourishing Traditions

You've done your homework and got the most basic but essential equipment to start your fermentation experiments; and you've also followed the steps to make your first batches of fermented vegetables. When you see that the pH sets at 4.6 or below, then there comes the "moment of truth": time to celebrate and open the jars to taste the flavorful fermented vegetables you made.

As you twist the lid to open the jar of fermented veggies, you may notice the bottle sizzling and bubbles starting to form, as if you opened a bottle of Coca Cola. This is expected as the fermentation process produces carbon dioxide, which brings up the air pressure inside the bottle; and opening up the jar releases the air inside and helps balance air pressure inside and outside of the bottles.

Because of this, it is a good idea to place a big bowl underneath the bottle, or let the bottle hover above the sink where the latter can hold the brine coming out of the jar as a result of air coming out, so that you can keep your countertop and kitchen floor clean.

Below the lid are the followers. You may want to remove them and place them on a clean spot, before you get the ferments out to taste. Then, after you take out the amount of fermented vegetables you want to consume, rinse the followers under filtered water before putting them back on the top of the ferments.

Then the fun begins – now you can officially start to enjoy what you prepared for yourself.

But, wait.

After I did a round of research, it turns out there is more to it than just opening the jar and eating the fermented vegetables. There are some basic guidelines you should follow and be aware of, too. Below I've summarized 12 tips to help you safely enjoy them, minimize your worry and reduce any surprises that you may not want. This should richen your experience.

1. Pay Attention to the Way You Take the Vegetables Out of the Jar.

You now know that you carry ten times more microbial cells than human cells inside the body. Microbial cells are everywhere, including in your tongue area. If you eat directly from the jar, you may run into the risk of introducing new bacteria into the batch, which can cause contamination. As an alternative and as a way to preserve the fermented foods, it is advised to take whatever amount you decide to have from the container with clean utensils, place them into a new bowl, and eat from that bowl.

2. Start Small and Slowly Add Up the Amount.

To train your gut to slowly adjust to the newly introduced environment, you can just start with a few spoonfuls of fermented vegetables to tease your gut and hint that more good things are coming up. Once you feel that the gut can adapt to the new foods with little reaction, you can then slowly increase the amount little by little all the way up to a quarter to half a cup a day. Always remember, giving the body and your gut a slow, gradual and steady transition is the key, so that the stomach will not get shocked.

3. Be Prepared for Potential Detox Symptoms.

Fermented foods can be powerful detoxifiers. As you introduce fermented vegetables into the body, they start killing off harmful bacteria inside the gut, which can lead to injured or dead bacteria releasing their toxins into the blood tissues faster than your body can comfortably process them. This can potentially cause a sudden immune response and exaggerated inflammation, causing Herxheimer Reaction (also known as die-off reactions, or detox symptoms).

Here are some examples of symptoms from die-off reactions: gastrointestinal distress, headaches, skin rashes, diarrhea, and skin problems particularly breakouts on the skin and chest, intestinal gas.

Fortunately, these symptoms cannot last for long, usually about two to three days (in rare cases, symptoms can last a few weeks); they just reflect that your body is adjusting to the new internal environment. My first detox reaction was mild itchiness on the back and elbows which lasted for quite a few days; but it was nothing unbearable. But now, I don't have any reaction after I consume fermented vegetables so long as I properly control the right amount I am consuming. That's a sign to me that my body has adjusted to the new dietary routine I have incorporated.

It is important to make small tweaks as you begin to taste fermented veggies. For example, when you are experiencing Herxherimer Reaction, instead of giving up altogether, you should just immediately cut down on the amount of consumption by half; also, cut down the frequency from daily consumption to every other day. Below is the exact quote from Caroline Barringer during an interview with Dr. Mercola:

'Don't easily give up and conclude that this is not for you: let your internal intelligence guide you, and if you see something or feel something that's not so right, don't dismiss cultured foods and say, 'oh, that was bad for me, it caused a reaction.' That's not what your body's telling you. Your body's telling you, 'slow down'.'[1]

4. Enjoy Fermented Vegetables Daily.

Ideally, if you don't experience die-off reactions, it is most beneficial to enjoy this kind of food at least once or two times a day. If your body can handle these foods with no problems at all, consider consuming them three times a day.

5. Remember, As Always, Only Take Small Amounts.

There is always a balance in everything. Just because fermented vegetables give you a substantial amount of health benefits, it doesn't mean that you should consume them without control. Although it is a general guideline that you start with very few and then add up the amount when your body adjusts to the change in diet. You should also keep very clear about the big picture: fermented vegetables should only take up a small amount of your total daily food intake.

As Sally Fallon, author of *Nourishing Traditions* says, "lactic-acid fermented vegetables and fruit chutneys are not meant to be eaten in large quantities but as condiments."[2] During my interview with Katz, he gave very similar advice that we should eat fermented vegetables in moderation only and that it is better to eat in small amounts consistently – rather than in large quantities – and that this practice gives you tremendous amounts of benefits down the road.

Zhihong Fan, member of China Nutritional Study Council and adjunct professor of Nutrition and Food Safety in the Department of Food Study in China Agriculture University, also advises that everyone consume only a small portion of fermented vegetables and consider them as a form of appetizer.

In my interview with Donna Schwenk, who is a strong believer of the effective healing power in fermented vegetables, she said that nowadays consuming ¼ to ½ a cup of fermented vegetables daily has become part of her routine and dietary maintenance plan to keep the gut functioning properly.

A lady I know from Vancouver told me that she loved having fermented foods and once over-ate them during a layover in Japan and ended up having a sick stomach throughout the whole flight. I had similar experience, too. Here is my lesson learned by consuming too much:

Two weeks after making my 16 bottles of fermented vegetables, I opened all of them to test their pHs and tasted a little bit of everything from all these ferments. I ended up developing a feeling of sickness in the stomach hours later during the day. Fortunately, that only lasted for less than 20 minutes, which could be a sign of my detox symptoms and suggests that I may have consumed too much at the same time.

As you see, to help the gut stay on the healthy side, you just need a small portion of fermented foods to cultivate beneficial bacteria. What's more, don't forget that a balanced diet with an overall healthy lifestyle is still your foundation to radiant skin and optimal health. This is very important to keep in mind, as people tend to over-consume foods without control after they learn that particular types of foods can bring benefits to the body. They forget that it is all about balance in everything. We should always keep this in mind.

6. Stay Persistent.

If you want long-term transformative results, you need to develop your perseverance and be patient with the progress. According to Caroline Barringer, it might take two and a half to three years to enjoy more stable and substantial benefits. But do not get discouraged just yet. She also thinks that initial results can be seen just by you consuming two to three servings.[3]

Overall, do not easily give up merely because you cannot see physical benefits immediately. Stick with the new dietary habits for at least one week and try feeling the subtle difference your body is experiencing.

7. Enjoy Fermented Vegetables of Different Kinds.

One reader used to ask: "what is the most beneficial fermented vegetable to start with?" The answer is there is not a particular fermented vegetable that is the most beneficial, as one type can carry different microbes than another. That's why it's recommended that you vary up the types of fermented vegetables you consume. But I do have some personal recommendations to help you get started based on the easiness of making them and their tastes. You can find more in Chapter 12.

Diversity matters. There are more than 10,000 microbial species in the human ecosystem. To help them thrive and do the work for you, you will need to introduce different types of microbes instead of just sticking to sauerkrauts only made of cabbage all the time.

Therefore, it is important to switch between different types of fermented vegetables to make sure that you get different bacteria and different strains within the same bacteria from different fermented vegetable sources. Eventually, you may also want to move onto other food groups other than vegetables to further diversify some microbial profiles from foods. Because this book primarily focuses on fermented vegetables, we will not go deeper into other types of fermented foods. But at least that gives you some food for thought.

8. Incorporate Fermented Vegetables in Your Daily Life.

One of the easiest things you can do with fermented vegetables is enjoy them as appetizers. Alternatively, you can mix them into your main meals.

In her book *Nourishing Traditions*, Sally Fallon says that fermented vegetables "can go beautifully with meats, fish of all sorts, and pulses and grains".[4] In *Cultured Food for Life* and her new book *Cultured Food for Health*, Schwenk inspires readers with creative ways to introduce fermented vegetables to your daily meals.[5] Take the examples of making kraut sandwiches and adding krauts into coconut milk fish soup. The Shockeys even inspire readers to make desserts out of fermented vegetables in their book[6] The point made here is to use your creativity to incorporate fermented vegetables into your daily life and make it fun.

As an example, one of my favorite ways to replenish microbes is to add a tablespoon of beet juice to my juice. Details of how I make my fermented beet juice and how I incorporate fermented vegetables into my daily life can be found in Chapter 13.

9. Do Not Forget Fresh Vegetables.

Some people ask, "If I consume fermented vegetables, does that mean I don't have to eat fresh vegetables; and are they the same?" I've come up with a detailed reason to explain the answer to this. To save your time, a quick answer for this is no. Below is my thought-out answer as to why fresh vegetables are *not* replaceable:

Most of section 1 is dedicated to showing you the importance of supplying the body with dietary fiber, which is partially derived from fresh vegetables, to feed the Permanent Residents inside us, or good microbes. Here comes the big picture once again: vegetables rich in dietary fiber are crucial for the PRs (who are the 100 trillion cells already inside us), whereas fermented vegetables are for the TCs (Transient Communities, the beneficial microbes we introduce externally).

You should still stick with what Michael Pollan suggests in his book *In Defense of Foods*: "[eat] mostly plants," because fresh vegetables have a lot of other major contributions besides feeding the microbes.[7]

They should form the staple of the diet. Vegetables are an important source of many nutrients such as minerals, vitamins, and dietary fiber to help with digestion, alkalizing the body, helping maintain an ideal environment that your body prefers (as mentioned above, your body likes a slightly alkaline environment to function properly), and reducing the risk of many diseases like heart disease, stroke, dementia, cancer, and arthritis.[8] For example, consuming green leafy vegetables can help fight diabetes;[9] loaded with magnesium, this type of vegetables can also help balance cortisol, your stress hormone.[10]

The amounts required for consumption are different, too. Unlike fresh vegetables, fermented vegetables should be enjoyed in small portions, especially for the salted fermented vegetables.

That's why it is a good idea to allow fermented vegetables and fresh vegetables to go hand in hand. This way, you will be able to enjoy benefits from both types to make sure that you have well-rounded nutritional intake while controlling your sodium level.

10. Yes, You Can Heat Up Fermented Vegetables.

But there are preconditions.

If you are not particular about what specific benefits you want to get and simply want to enjoy healthy foods, heating up fermented vegetables is doable. According to Katz, "the pre-digestion, nutrient enhancement, and detoxifying actions of fermentation can be of nutritional benefit whether or not foods are cooked after fermentation."[11]

On the other hand, if you are particular about preserving live cultures within fermented vegetables and would like to specifically enjoy the healing benefits from these live organisms, then heating up the fermented vegetables is *not* recommended. That's because these live cultures cannot survive in heat exceeding around 115 °F (47 °C).

11. Be Creative about Using Brine.

If you think about it carefully, the brine in fermented vegetables is almost the same as fermented vegetable juices. The only difference is that they contain different amounts of sodium. Therefore, it is totally okay to drink shots of brine as a digestive tonic, as long as you make sure that you keep your sodium intake in check.

Since there are live cultures in the brine as well, you can also use them as starter cultures for your next batch of vegetable fermentation. I once used the brine from fermented cabbage as the brine for fermenting chopped cabbage leaves and carrots; this helped enrich the flavor of the whole ferment and, since I like sour foods in general, I enjoyed the extra kick of sourness brought up by this brine.

Alternatively, you could also use the extra brine for salad dressings, which is what I do from time to time. Or, if you are not particular about preserving live cultures in the brine, you can also use it as a base for soups to enhance the flavor.

Just as there are a myriad of ways to incorporate fermented vegetables into your daily life, there are many creative experiments you can do with the brine. The examples laid out in this book will serve as your inspiration and help you extend your imagination to a further place. Be sure to share your experiments with me, too. You can find more details on how to stay connected with me in the "About the Author" section at the end of the book. I'd love to learn from you.

12. Wait for a Little While.

Here is one final tip I have for you. After the taste reaches your ideal flavor, you may transport them into the fridge (more details follow in Chapter 8). I have learned that when fermented vegetables sit in the fridge for one week or longer, they generally taste even crunchier and have more flavors, too.

My guess for why is that the reduction of moisture in the fridge is what crisps them. And, I was surprised to find out that my experience was later validated by a quote I discovered in the book *Nourishing Traditions*: "the fermentation of the aromas don't come out until a later stage, during storage – a way to ennoble fermented vegetables."[12]

Interestingly, Schwenk shares a familiar lesson in her book *Cultured Food for Health*, too: "I find that many of my vegetables taste better after six weeks in the fridge. It's fun to taste your vegetables at different stages to find out when you like them best."[13]

By keeping all the above in mind, you will be able to enjoy consuming fermented vegetables safely and reap the most benefits out of it, while having fun at the same time. Of course, since you are not going to finish all you make in just one sitting, then how should you preserve them properly? It turns out that there is a set of guidelines that you need to know, too, which I will talk more about in the next chapter.

Chapter Recap

- *Use clean utensils to take out ferments from the jar to avoid contamination.*
- *Start with eating only a spoonful of your favorite fermented vegetable; then slowly increase it to ¼ to ½ cup a day.*
- *If you experience detox symptoms, cut down the amount by half or the frequency from daily to every other day.*
- *Ideally, you can consume fermented vegetables at least once or twice a day.*

STEP FOUR – EAT FERMETNED VEGETABLES THE RIGHT WAY

- *Always consume fermented vegetables in moderation.*
- *Staying persistent in consuming fermented vegetables is the key for substantial results.*
- *Diversity matters. Make sure that you try different types of fermented vegetables.*
- *There are many ways you can enjoy ferments: eat them as an appetizer; or, mix them into your meals. You can find my recipes on how to blend fermented vegetables into your meals in Chapter 13.*
- *Fermented vegetables can never replace fresh vegetables.*
- *You can heat up fermented vegetables and still enjoy some of their nutritional benefits; but heating them up can kill microbes when they are exposed to heat exceeding 115 ºF (47 ºC) or more.*
- *The brine is drinkable as long as the salt content is 3% or less. You can enjoy it as a digestive tonic, as starter cultures, as salad dressings, or as a base for soups.*
- *After sitting in the fridge for a week or longer, fermented vegetables taste crunchier and develop more flavors.*

Exercises

- *Each time I take ferments out of the jar, I need to use _____ to take them out to avoid contamination.*
- *When my bottle of fermented vegetables is ready to serve, I know that I may expect some detox symptoms after I consume them, which may include _____ ___. That's why it's important for me to start with _____ a day only, or to cut down consumption to every other day.*
- *Brine is drinkable as long as the salt content is _____ or less.*

Checklist

- *Visit www.tracyhuang.me/fvm-resources for this checklist: "12 Ways to Enjoy Fermented Vegetables Safely".*

Chapter 8:
Step Five –
Store Them Well

"The sign of successful lacto-fermentation is that the vegetables and fruits remained preserved over several weeks or months of cold storage."

– SALLY FALLON & MARY ENIG
Nourishing Traditions

When the ferment is ready to serve, it is time to transport the bottle to the fridge to slow down the fermentation process.

If you use cabbage leaves as followers, you may find that leaves wilt during fermentation process and can no longer effectively keep ferments submerges underneath the brine. In that case, you can discard these cabbage leaves and replace them with one or two fresh cabbage leaves as new followers to keep ferments submerged, before you store the bottles into the fridge.

It is hard to say how long a jar of ferments can last, as it depends on many factors such as the type of vegetables, saltiness in the brine, the temperature the jar is exposed to, and other variables. But, overall, as long as the ferments still taste fresh and crunchy and don't smell bad to you, you can still enjoy them.

As a general rule of thumb, it is good to store ferments for anywhere from six to 12 months. Yet, in reality, if you are making fermented vegetables at home and are just getting started, chances are you will be using small containers to make them. Usually they are one-quart, one-pint, and half-pint jars. With jars of these sizes, it is likely that you can finish all the ferments within just a few weeks, or even in a few days if you share them with your family. If you store them in the fridge, they should be able to last for that long.

At the same time, it is actually better if you can finish up the bottle of ferments as soon as possible. Studies show that "since the brining process involves preservation as a result of, or accompanied by, microbiological activity, nutrients may be lost through utilization by microorganisms"; therefore, the longer you wait, the more likely it is going to lose nutrients like vitamins, protein, and minerals.[1]

After ferments stay in the fridge for a while, you may find that the brine "evaporates". That happened to one of my experiments, when I reached out to Shockey from *Fermented Vegetables* for help. She explained that it shouldn't be a big concern as long as: a) ferments are tightly packed; and b) you are going to finish them very soon. Here is one final note: most jar lids are metal. If you want to be able to store ferments for a few months or longer, you can consider covering wax papers on the top of the jar before screwing on the lid or ring. This keeps rust from forming on the inside of the lid or along the inside of the ring.

After you learn how to make, consume, preserve fermented vegetables in the right way, what is next? It's time to take actions. In the next chapter, I would like to discuss action steps you can take to help you get further into the world of fermentation and start having more fun.

Chapter Recap

- *You can consider transporting the bottle of ferments into the fridge once they are ready to serve.*
- *If cabbage leaves you use as followers wilt during fermentation process, you can replace them with fresh ones to keep ferments submerged.*
- *Generally, ferments can last from six up to 12 months; but it's best if you can finish them as early as possible to ensure you get the most nutrients out of them.*
- *It is okay if the brine does not fully cover the ferments as long as: a) they are tightly packed; and b) you plan to finish them in a very short time.*
- *If you plan to keep ferments for a longer period of time, you could apply wax papers on the top of the rim and right underneath the lid to minimize chances of rust on metal lids due to long storage.*

Exercises

- *The ready to serve ferments usually can last _____.*
- *After the ferments are sitting in the fridge for a while, it should not be a concern if I see they are not submerged as long as _____.*

Checklist

- *Visit www.tracyhuang.me/fvm-resources to check out this checklist: "Simple Steps to Enjoy Fermented Vegetables after You Open the Jar".*

Chapter 9:
Step Six –
Go Beyond Knowledge

"You may never know what results come from your action. But if you do nothing, there will be no result."

– MAHATMA GANDHI

I hope by now you have cleared up your worries and are well prepared mentally to get started. The purpose of this book has been to help you get over your mental hurdles and break down the fermented process into easy and manageable steps, so that you can quickly experience the benefits of vegetable fermentation and explore a whole new territory, where you can even implement alternative solutions to heal the body.

Knowledge itself will not give you transformative results until you act on it. Why not give yourself an opportunity to welcome a new, safe, and cost-effective way that can transform your health? Below I'd like to share a piece of meaningful advice from Lisa Heldke, professor of philosophy at Gustavus Adolphus College, quoted from *The Art of Fermentation*:

> *"Rules alone cannot teach you how to do something. Written instructions are insufficient. Culturing food reminds me of this regularly. You can't learn it all from reading—whatever "it all" is. Set aside the book and "read" the ingredients; pay attention to them as if they matter. Put down the instructions and ask a human being for advice. Be prepared to be foolish, ignorant, naive, and wrong. Be grateful for the insights you are given—and don't complain because the teacher rambled."*[4]

Start from today spending a few minutes jotting down necessary tools you need, pulling out one or two recipes you would to try first from this book, and listing all ingredients on a piece of paper. Then, find a weekend to go online shopping or hit local grocery stores to gather all you need before you start your first fermentation experiment in the kitchen. Since included in this book are all established recipes designed by industry experts with minor tweaks from me and personal commentaries, you can feel assured that they are tested, worry-free, and definitely worth a try. Once again, you can get access to recipes I used for my own experiments by visiting Chapter 12, and to how you can incorporate ferments into your daily life in Chapter 13.

Additionally, I encourage you to explore local workshops you can join for free or at a modest fee; or, you may check out other people's stories on their fermentation journeys. You may feel a bit lonely as you think you have to do this all by yourself. However, when you look around, you will discover there are many people who are already doing this; and you will also find out that fermented vegetables actually have been benefiting them and their family for quite a while now.

Joining a community makes your experience more fun. You can start by searching on Meetup.com, public Facebook groups hosted by local communities, and public Facebook events. I sometime found fermentation workshops in local libraries, too. By actively searching, I even found a free festival – Boston Fermentation Festival – hosted by a local non-profit organization – Boston Ferments – dedicated to fermentation education where I got to connect with tens of thousands of people and amazing authors, too.[2]

In the meantime, I look forward to seeing pictures of vegetables you choose to ferment, the process of you making them, and what they eventually look like. You can connect with me via Facebook, Twitter, or Instagram, and use the hashtag "#fvm_fun" to let me know about all these moments, and what you learned along the way. Details of my social media channels can be found at the end of this book in the "About the Author" section.

I welcome you to send me more questions, so that I can continue to be of assistance to address your concerns. If you haven't done so yet, you're also invited to subscribe to my newsletter at www.tracyhuang.me/join to stay updated with my continual learning journey on fermentation and many other explorations on how to combine traditional wisdom and food science to maximize our health and wellbeing.

Chapter Recap

> •Here are some ways you can take actions after finishing this book: a) jot down necessary tools you need; b) pull out one or two recipes from Chapter 12 to start your experiment; c) list all ingredients you need to prepare; d) plan a weekend for grocery shopping; e) explore local workshops; f) check out other people's stories on their fermentation journeys; and g) join a local community by visiting Meetup.com, related public Facebook groups, or public Facebook events.

Exercise

- *One thing I plan to do after finish this book is* _____.

SECTION III: VEGETABLE FERMENTATION DISCLAIMER AND FINAL THOUGHTS

Chapter 10:
Oops… Vegetable Fermentation Is Not the "Holy Grail"

"Finding a probiotic that is beneficial for you requires that you become a microbiota experimentalist. It is best to test a variety of probiotics to determine what works for you."

– ERICA SONNENBURG & JUSTIN SONNENBURG
The Good Gut

Congratulations on having gone this far by now! I hope your journey so far has been smooth and rewarding. I tried my best to pave the way for you and make sure that all bricks are laid in alignment, so that you can walk on them feeling assured, safe and confident:

Your body is not just about you, but also the trillions of microbes who are on and under your skin and who fight for your own health (Chapter 1). Because of this, you need to upgrade your current modern yet not-so-microbe-friendly ways of living, as they are threatening their existence as well as your health (Chapter 2).

Compared to before reading this book, you are now observing with a 3,000-foot view on how vegetable fermentation is tied into a bigger picture – there are two types of microbes we need to feed the body: one is called "Permanent Residents" which need to be fed with dietary fiber-rich foods; and the other is called "Transient Communities" which you get through introduction of probiotics supplements or fermented foods (Chapter 3). Hence, you have learned that consuming fermented vegetables is a piece of the puzzle. To make sure that this piece is put in the right place, I have taken you through six steps to help you minimize mistakes (Chapters 4 to 9).

I have shown you the woods and taken you into the trees. Now I am leading you out of the woods, away from the trees again by telling you this depressing yet not-so-depressing fact: vegetable fermentation is not the "Holy Grail".

Due to the exclusivity of microbiota from every individual, different people may react to the same type of fermented foods differently – the microbes you introduce may or may not have synergy with the microbes that are already residing in you. In other words, you may not be able to find the same miraculous effects claimed on websites you have read or people you have learned from. You may even find that your body is currently allergic to fermented vegetables due to your having histamine intolerance (which happened to some people I surveyed).

I am not saying this to be a party pooper. I am saying this because I believe that it is always good to have a big picture in mind – hope for the best; but prepare for the worst, anyone? I am saying this to let you know that circumstances may happen but there is still much that you can do as long as you stay sane.

The most important message in this book is to help you become aware of the importance of nourishing and cultivating microbes you are carrying. If you are fully aware of the ways you are currently killing your friendly and capable allies, and you take actions to keep those activities in control, then this book has done its job. If you have made a decision to turn to consuming more dietary fiber rich foods to help feed the warrior microbes you are living with, then this book has done its job, too.

But this doesn't mean that you can discard the whole six chapters I have written for you, guilt-free, on how to make and enjoy fermented vegetables. It is not because I want you to feel guilty for me, but because I don't want you to be guilty for not taking the chance to explore further opportunities that can improve your health by taking advantage of vegetable fermentation. Yes, detox symptoms may happen to you in a mild way (but it may not). Don't give up so easily. I recommend you stick with a routine to see if consuming fermented vegetables works for you or not. Start with a very small amount, perhaps, one tablespoon. Patience is a virtue.

And, lastly, which is the most exciting in my opinion – build yourself a system to find out what is working for you and your microbes; and cultivate a mindset of you becoming your own experimentalist, because you are own best doctor and know what is best for yourself.

To do this, you create for yourself a specific plan or schedule to try out different kinds of fermented vegetables on a systematic scale: first, by introducing only one type of fermented vegetable slowly; second, by slowly adding up the amount and consuming them regularly; third, by adding up the frequency of consumption to a daily level, and being persistent for at least a week to feel the different effects.

To perform this experiment successfully, it is important to stick with one particular group of fermented vegetables first, before you move onto the next group. I got this idea from the Sonnenburgs, authors of *The Good Gut*, and really like it.

Each type of fermented vegetable has its own charm that your microbes may find synergy with. It is like you being single and looking for a soulmate: you wanted to expose yourself in different social events like bars, potluck gatherings, movie nights, baby showers, parties, or friends' weddings, as you would never know who you would meet that you then decided to not part with for the rest of your life. As always, stay in the unknown; yet always anticipate that something magical will happen.

Chapter Recap

- Be prepared that, due to exclusivity of individual microbiota, you may not be able to find the same miraculous effect claimed on a website or by other people you learned from.
- But do not get discouraged. There are things that you can do to: a) acknowledge the importance of microbes, minimize microbe-killing activities, and feed microbes with dietary fiber-rich foods; b) be patient with detox symptoms; and c) build yourself a system to experiment on one group of fermented vegetables at a time.

Exercises

- The most important message I need to take away from the book is _____. Regarding that, one action I need to take for the sake of the health of myself and my microbes is _____.
- I plan to start with experimenting _____ (fill in a specific type of fermented vegetables you are thinking of starting now). I will build myself a system to explore whether this particular type of fermented vegetable is working for me. And my plan looks like this _____.

Chapter 11:
Final Thoughts

"(Traditional) food wisdom is worth preserving and reviving and heeding."

– MICHAEL POLLAN
Food Rules

Having just come off the last chapter, you've just learned the most important message you should go home with: to start preserving and cultivating your gut flora from today onward. No further delay.

The study of the recent exciting discoveries regarding microbiology helps me redefine "bacteria" we inevitably face every day in a different way. I used to think bacteria were all harmful and hand sanitizer was one of my going-out essentials. I now have learned that our civilized society may have prompted us to live in an overly protected environment where everybody is obsessed with ultra clean environments, such as excessive use of anti-bacterial soaps and sanitizers which can kill off even the beneficial bacteria that are covering our bodies.

A civilized society also has brought along convenient processed comfort foods, chronic stress from work and other things in life that together gradually wipes out our helpful bacteria inside the gut, leading to a weakened body and all types of symptoms. It turns out that we humans are not as powerful or independent as we think, as we need to work with the bacteria. They are our allies to sustain life, proven by how our bodies contain 10 times more of their cells than our own human cells. My take-away lesson is we should stay humble and learn to be kind to our bacterial roommates.

The future of microbiologic studies is emerging. Right now, there is the American Gut Project to help you look into your gut and microbial profile to analyze your health status. It is even predicted that, in the future, microbiotic tests will become just as important as blood tests, as something to be checked regularly as a way to monitor your health.

Because of this, vegetable fermentation has become more than just a fun experiment you do in your kitchen-lab, but it also marks the beginning of your realization of the importance of maintaining gut health by cultivating gut flora; it indicates that you are part of the movement, getting ready for something potentially more exciting to come.

Meanwhile, since there are over 10,000 microbial species of bacteria in the body and lactic acid bacteria are the primary bacteria produced from vegetable fermentation, only one type of food fermentation is not enough. Gradually, we all need to branch out to other types of fermented foods to diversify our microbial profiles. Microbial diversification is very important for gut health.

Then, we may even need to go beyond food and think at a lifestyle level to explore more ways to preserve and cultivate microbes. Take the examples of avoiding the excessive use of antibiotics, going out to the field to play with dirt and soil more often (such as in the form of gardening), re-thinking about the need to pursue excessive cleanliness (such as the use of anti-bacterial soap each time you finish your business in the bathroom). Just as the study of the importance of microbes is at its beginning stages, we should keep on exploring ways to preserve and cultivate gut flora, as well.

Next, I also want to talk about the importance of combining food science with ancient tradition.

I have talked about recent discoveries in microbiology in Chapter 1 and a revival of the ancient practice of fermentation in Chapter 3; we have also covered that it is scientifically proven to be vital to maintain gut health because our gut functions like a second brain, which coincides with the fact that Traditional Chinese Medicine has been treating the stomach and spleen as crucial organs for optimal health and well-being for over 2,000 years. These suggest to me the synergy between ancient wisdom about healthy eating, and using science to perfect ancient practices.

In other words, the rising attention to fermentation and the increase of fermented food advocacy is a sign that we should continue to respect ancient teaching while leveraging science to improve our health in a more precise way.

Traditional wisdom, though still needing more scientific study to fully understand its impacts, is something that has been preserved and practiced successfully on real human bodies for hundreds of thousands of years. I think we should keep an open mind to constantly look to the past for ideas and suggestions on how to eat, like the practice of fermenting foods. Indeed, as Michael Pollan points out, you don't have to become scientifically savvy to enjoy the benefits of real foods; and much of traditional food wisdom is worth looking back to as your important point of reference for health and wellness.

This is not anti-science. On the contrary, I also believe that science can provide us with powerful and effective ways to meet our body's needs, to understand foods, and to help us learn how to better consume them. In the context of vegetable fermentation, food science has helped us make conscious decisions in the preparation process to control sodium intake within the suggested range and to ensure the pH level is under control with the use of pH test papers or meters. Through the process of science reviving ancient wisdom, science has offered us a safer way to follow tradition. Because of this, we should continue to leverage science.

That's why I personally think that health and food movements should take advantage of both tradition and science to reap the most benefits.

Chapter Recap

- *It is important to start cultivating your gut flora from today onward. You can start from vegetable fermentation, gradually moving onto consuming other types of fermented foods, then expanding your horizon beyond foods to cultivate microbes from a lifestyle level.*
- *It is important to combine science and traditional food wisdom to reap the benefits of food.*

Exercises

- *To nourish microbes, I should not limit myself to just fermented _____. Rather, I need to broaden my horizon to other types of fermented _____ and even the overall lifestyle.*
- *A combination of _____ and _____ can help me reap the most benefits from foods.*

SECTION IV: EXPERIMENTS AND RECIPES

Chapter 12: Sixteen Vegetable Fermentation Experiments

"Every artist was first an amateur."

– RALPH WALDO EMERSON

Below are 16 experiments I performed. As I was making them at home, the temperatures fluctuated between 53 °F (around 12 °C) and 68 °F (20 °C).

My top two recommendations are salted white cabbage sauerkraut and beet kvass because each only needs two ingredients, can be made in 15 minutes or less, and can be ready to serve in around three to four days. Most importantly, they both taste great.

If you especially want to make crunchy fermented vegetables, I would also recommend you start with eggplant garlic kraut, cucumber pickles (with tannins), bok choy pickles, Napa cabbage kimchi (presoaking method), assort root kimchi, asparagus kimchi, or ferment red cabbage (with a regular lid or with an airlock).

Kitchen Experiments at a Glance

1. Salted White Cabbage Sauerkraut
2. Salt-Free White Cabbage Sauerkraut (with Celery Juice)
3. Sweet Onion Spinach Kraut
4. Eggplant Garlic Kraut
5. Leak Paste
6. Fermented Baby Kale
7. Beet Kvass
8. Cucumber Pickles (with Tannins)
9. Cucumber Pickles (without Tannins)
10. Bok Choy Pickles
11. Napa Cabbage Kimchi (Kraut Making Method)
12. Napa Cabbage Kimchi (Presoaking Method)
13. Assorted Root Kimchi
14. Asparagus Kimchi
15. Fermented Red Cabbage (with a Regular Lid)
16. Fermented Red Cabbage (with an Airlock)

Detailed Experiments

1. Salted White Cabbage Sauerkraut

Time: about 15 minutes

Yields 1 pint

I followed the kraut making method in this experiment.

By mixing salt with the shredded cabbage leaves and letting them produce their own brine to keep them submerged, I wanted to see if the salted ferments would taste different than the unsalted ones.

In this experiment, I followed the guidelines from *Real Food Fermentation* by Alex Lewin.[1] After making and tasting this kraut and salt-free sauerkraut made with celery juice (which I will describe in the second experiment), I learned that sauerkraut with sea salt could make ferments taste crunchier and more flavorful.

For this particular batch, its pH already dropped to below 4 by Day 4. Would I recommend making salted white cabbage sauerkraut for beginners? Of course. And, it is easy to make.

Ingredients

- 1 pound and 1.5 ounces white cabbage
- 2 teaspoons sea salt

Instructions

- Discard the outer leaf if the cabbage is not organic and then keep one or two whole leaves as the followers later on.
- Cut the cabbage into thin slices.
- Add salt and massage for about 8 minutes.
- Pack everything into the jar and press firmly until you see the ferments are submerged underneath the brine, and keep 1-2 inches of distance between the rim of the jar and the top of the brine.

- Roll up the 1-2 leaves saved at the very beginning and fit the followers into the jar (make sure that you firmly press down followers against the inner wall of the jar, so that they can be "stuck" at a particular position. Also, keep the followers submerged in the brine as well).

- Place a lid on the top and turn it to seal the jar until it just stops turning.

- If necessary, add a sticky note to remind yourself of the ingredients, the date of making the ferments, and the pH values.

- Store the jar in a container in case of brine leak.

- Closely monitor the whole process especially in the first few days when the fermentation process is most active: the brine may leak and the ferments may float to the brine surface and be exposed to oxygen (if this happens, you will need to press down the kraut and the followers to keep them submerged).

2. Salt-Free White Cabbage Sauerkraut (with Celery Juice)

Time: about 10 minutes

Yields ½ quart

You will need a 32-ounce jar to begin with because of little osmosis at the beginning.

In this second experiment, I also used the kraut making method. Instead of adding salt, I added celery juice to form the brine. I learned this trick from Dr. Mercola's interview with Caroline Barringer, the certified Nutritional Therapy Practitioner.[2]

Not only did I find that salted sauerkraut tasted better, I also discovered that the salt-free version could get mushy easily although both sauerkrauts had been fermented for the same amount of time. Therefore, it helps if you could start consuming the salt-free version as soon as the pH of the ferments reaches 4.6 without waiting further, so that you can enjoy them as crunchy as possible.

Ferments can be ready in about four days. I would recommend it for beginners, if you believe you are already consuming too much salt these days and would like to keep your sodium intake in check; you may also consider a salt-free version simply because you want to see what it tastes like.

Ingredients

- 1 pound and 1 ounce white cabbage
- 7 celery stalks (about 10 ounces)

Instructions

- Discard the outer leaf if the cabbage is not organic and then keep one or two whole leaves as followers later on.
- Cut the cabbage into thin slices.
- Pack the sliced cabbage leaves into the jar and use a tamper (or its alternative) to push them down hard.
- Juice the celery stalks and pour the liquid into the jar.
- Roll up the 1-2 leaves saved at the very beginning and fit the followers into the jar, and keep 1-2 inches of distance between the rim of the jar and the top of the brine (make sure that you firmly press down the followers against the inner wall of the jar, so that they can be "stuck" at a particular position, and also keep the followers submerged in the brine as well).
- Place a lid on the top and turn it to seal the jar until it just stops turning.
- If necessary, add a sticky note to remind yourself of the ingredients, the date of making the ferments, and the pH values.
- Store the jar in a container in case of brine leak.
- Closely monitor the whole process especially in the first few days when the fermentation process is most active: the brine may leak and the ferments may float to the brine surface and be exposed to oxygen (if this happens, you will need to press down the kraut and the followers to keep them submerged).

3. Sweet Onion Spinach Kraut

Time: about 13 minutes

Yields 1 pint

Krauts are more than just sauerkraut. In order to learn more about fermenting vegetables other than cabbages, I made two krauts to play around with different vegetable combinations. In this experiment, I mixed leafy greens with juicy and mildly crunchy sweet onions.

This following recipe was inspired by *Fermented Vegetables* by Kirsten and Christopher Shockey.[3] For this particular batch, its pH of ferments dropped to 4.41 by Day 6. This shows it takes about a week before the sweet onion spinach kraut is ready to serve.

The fermentation of the chlorophyll-rich spinach may generate smell that not everybody will like. I personally feel that the taste and smell was too strong for me. Since you are not sure whether you will like the taste and smell, I would recommend you start your vegetable fermentations with other recipes that I highly recommend first. But if you are really adventurous, then give this recipe a try.

Ingredients

- 12 ounces spinach
- ½ sweet onion
- 1 teaspoon sea salt
- 0.25 ounce oregano
- 1 tablespoon lemon juice
- 1 or 2 red cabbage leaves

Instructions

- Chop the onion into pieces.
- Add salt and massage for about 3.5 minutes.

- Add other ingredients and continue to massage for another 5 minutes (add one small portion of spinach each time).
- Pack everything into the jar and press them down firmly.
- Make sure that the ferments are submerged in the brine and there is a 1-2 inches distance between the rim of the jar and the top of the brine.
- Cut one or two small pieces of the red cabbage leaves, so that they can fit into the jar.
- Place them on the top of the brine, and firmly press down the leaves, so that they are submerged in brine, too.
- Place a lid on the top and turn it to seal the jar until it just stops turning.
- If necessary, add a sticky note to remind yourself of the ingredients, the date of making the ferments, and the pH values.
- Store the jar in a container in case of brine leak.
- Closely monitor the whole process especially in the first few days when the fermentation process is most active: the brine may leak and the ferments may float to the brine surface and be exposed to oxygen (if this happens, you will need to press down the kraut and followers to keep them submerged).

4. Eggplant Garlic Kraut

Time: about 14 minutes

Yields ½ pint

This is another kraut variation where cabbage leaves are not involved. I specifically chose vegetables with different textures – in this case, the soft eggplant was mixed with the crunchy carrots – to find out what the combination would taste like.

The recipe is again inspired by *Fermented Vegetables*.[4] Eggplant keeps a lot of moisture, making it easy for the brine to form. In addition, basil and garlic mixed very well with each other and added very pleasant smells and flavors to the ferments. To my surprise, the soft eggplant chunks later turned out to be crunchy when the whole batch was ready to serve. I guess it was because of the tannins (in the black tea leaves) which enhanced the crunchiness of the eggplant. According to my experiment, its pH dropped to 4.16 by Day 3. And, I also recommend this to beginners.

Ingredients

- 6 ounces eggplant
- 1.4 ounces carrots
- 1 teaspoon sea salt
- 3 garlic cloves
- 0.2 ounce basil
- 1 tea bag black tea leaves
- 1 or 2 cabbage leaves

Instructions

- Chop the eggplant and carrots into small chunks.
- Mince the garlic cloves and add the garlic to the mixture.
- Add salt and massage for about 3-4 minutes.
- Place half of the basil leaves at the bottom of the jar.
- Place the black tea leaves from 1 tea bag and pour them into the bottom of the jar.
- Put in half of the mixture (eggplant and carrot chunks and minced garlic).
- Put in the rest of basil leaves.
- Put in the rest of the mixture.
- Cut 1-2 small pieces of the cabbage leaves and squeeze them in on the top of ferments as the followers.

- Make sure that the ferments and the followers are submerged with 1-2 inches of space between the top of the brine and the rim of the jar.
- Place a lid on the top and turn it to seal the jar until it just stops turning.
- If necessary, add a sticky note to remind yourself of the ingredients, the date of making the ferments, and the pH values.
- Store the jar in a container in case of brine leak.
- Closely monitor the whole process especially in the first few days when the fermentation process is most active: the brine may leak and the ferments may float to the brine surface and be exposed to oxygen (if this happens, you will need to press down the kraut and the followers to keep them submerged).

5. Leek Paste

Time: about 10 minutes

Yields 1 pint

Because pastes and bases are used as condiments, this particular kraut can be used to enhance flavors of a dish. Usually, more percentage of salt is considered when you make pastes and bases. I followed the introductions in *Fermented Vegetables* to make this to see what krauts with higher salt content would taste like as condiments.[5]

Leek is a form of leafy green. I was worried if it would generate strong smells like spinach. It turned out that the taste and smell of leek was pleasant. I would also recommend this to beginners. And leek paste can be ready in around four days. If the leek does not generate enough brine, you may need to add more from the outside; you can refer to my lessons learned from my sauerkraut making experiment (Chapter 6) to review the three ways to add brine from the outside.

Ingredients

- 12 ounces leek

- ½ teaspoon sea salt
- 1 or 2 red cabbage leaves

Instructions

- Chop the leek into pieces (you may use a blender for help).
- Add salt into the leak paste and massage for about 8 minutes.
- Pack everything into the jar and press down the ingredients firmly.
- Cut 1-2 small pieces of red cabbage leaves, so that they can fit into the jar.
- Place them on the top of the brine and firmly press down the leaves, so that they are submerged in the brine, too.
- Make sure that the ferments and the followers are submerged with 1-2 inches of space between the top of the brine and the rim of the jar.
- Place a lid on the top and turn it to seal the jar until it just stops turning.
- If necessary, add a sticky note to remind yourself of the ingredients, the date of making the ferments, and the pH values.
- Store the jar in a container in case of brine leak.
- Closely monitor the whole process especially in the first few days when the fermentation process is most active: the brine may leak and the ferments may float to the brine surface and be exposed to oxygen (if this happens, you will need to press down the kraut and the followers to keep them submerged).

6. Fermented Baby Kale

Time: about 3 minutes

Yields ½ pint

Although I am aware that chlorophyll rich kale may generate characteristic smell that not everybody will like, I still decided to try this experiment by making a small batch. I chose a half-a-pint jar and prepared the amount of kale enough to fit into this size. I performed this experiment also to try the whole leaf fermentation and see how this would turn out.

Again, I referred to the instructions in *Fermented Vegetables* as I made this.[6] As expected, this bottle of the ferments generated smells and flavors that were too strong for me. And the mushy texture may not appeal to everyone.

In addition, the pH of the fermented baby kale dropped more slowly than other types of fermented vegetables. In my experiment, the pH reached 5.93 after 22 days (I conducted the same experiment for three times and obtained similar results: the pH did not go below 5 after the baby kale had been fermented for 15 to 22 days). If the fermentation is too slow, the quality of the fermented vegetables can be compromised, because they cannot quickly create an environment that is acidic enough to inhibit potential pathogen growth.

The reason for the slow fermentation can be that there is less sugar in leafy vegetables, according to Breidt; one way he recommends is to mix leafy greens with vegetables that contain more sugar content like cabbage, instead of using only leafy greens.

For the above reasons, this is my least recommended experiment for beginners.

Ingredients

- 6 ounces baby kale
- ½ teaspoon sea salt
- 1 or 2 red cabbage leaves

Instructions

- Place the kale in a big bowl and sprinkle salt on the top.
- Massage for 1 minute (you can quickly see the kale leaves are wilted and the brine is formed).

- Pack the wilted kale leaves into the jar, and press them down firmly to keep the ferments submerged.
- Cut 1-2 small pieces of the cabbage leaves (so that they can fit into the jar) and squeeze them in on the top of ferments as the followers.
- Make sure that the ferments and the followers are submerged with 1-2 inches of space between the top of the brine and the rim of the jar.
- Place a lid on the top and turn it to seal the jar until it just stops turning.
- If necessary, add a sticky note to remind yourself of the ingredients, the date of making the ferments, and the pH values.
- Store the jar in a container in case of brine leak.
- Closely monitor the whole process especially in the first few days when the fermentation process is most active: the brine may leak and the ferments may float to the brine surface and be exposed to oxygen (if this happens, you will need to press down the kraut and the followers to keep them submerged).

7. Beet Kvass

Time: about 5 minutes

Yields 1 quart

This is my experiment for making fermented vegetable juice. I made my first fermented beet juice by following the instruction in *The Art of Fermentation* by Sandor Katz.[7] You don't need to add followers in this fermentation as the weight of the beet chunks is enough to keep the ferments submerged in the brine. As beets contain a lot of sugar, the initial stage of fermentation can be very active. Be sure to watch out for the potential brine leak.

I *highly* recommend you pick this as one of your early experiments, as beet kvass is very easy to make and can be ready to serve in just three days (in my experiment, the pH dropped to 4.00 by Day 3).

Ingredients

- 2 small beets (about 5 ounces)
- A few pinches Himalayan pink salt
- Filtered water

Instructions

- Peel the beets and cut them into small chunks.
- Add in a few pinches of Himalayan salt and mix well with the beet chunks.
- Fill the filtered water all the way to the top and leave 1-2 inches of space between the top of the brine and the rim of the jar.
- Use a clean stick, chopstick, or other long utensil to stir well.
- Place a lid on the top and turn it to seal the jar until it just stops turning.
- If necessary, add a sticky note to remind yourself of the ingredients, the date of making the ferments, and the pH values.
- Store the jar in a container in case of brine leak.
- Closely monitor the whole process especially in the first few days when the fermentation process is most active (based on my experiment, if you make sure that there is 1-2 inches of space between the brine surface and the rim of the jar, there should be no brine leak).

8. Cucumber Pickles (with Tannins) and 9. Cucumber Pickles (without Tannins)

Time: about 5 minutes

Yields 1 quart

I wanted to learn the difference in tastes between cucumber pickles with and without added tannins and, therefore, designed two experiments to check out the difference. The ingredients and steps of making both types of pickles are basically the same except that one contains tannin-rich black tea leaves and the other one doesn't.

After making my pickles, I realized that the number of cucumbers could have been more. Because I added only three baby cucumbers to each jar, this left room between cucumbers and caused them to float to the brine surface often. Therefore, I recommend you add a few more cucumbers to each jar to secure their places inside the brine by packing them tight and right next to each other.

Did the cucumber pickles with added black tea leaves enhance the crunchiness of ferments? Yes. So, don't hesitate to add some black tea leaves as you make pickles to enhance the texture. Besides black tea leaves, you could also consider horseradish, grape leaves and oak leaves.

By Day 4, the pHs of both batches already dropped to around 4.6. The pickles turned to be less salty than the commercial pickles. I may consider adding up the amount of salt next time when I prepare the brine to add a stronger flavor to cucumber pickles. On the other hand, the light salty flavor made the brine very tasty and drinkable.

The recipe below is inspired by *Real Food Fermentation* by Lewin.[8]

Ingredients for each bottle

- 3 baby cucumbers (about 7.7 ounces)
- 0.7 ounce dill
- 5 bay leaves
- 10 whole garlic cloves
- Brine (to make this brine, mix 6 cups of filtered water mixed with 1 tablespoon of sea salt; that's about 2.5 cups of the brine needed for each jar)
- 1 teabag black tea leaves

Instructions

- Cut off the cucumber blossom ends to avoid softening of pickles.
- Place the dill, bay leaves, whole garlic cloves, and leaves from the teabag (for one bottle only) into the bottom of the jar.

- Place the cucumbers vertically into the jar.
- Carefully pour the brine into the jar all the way to the top.
- Make sure that the ferments are submerged with 1-2 inches of space between the top of the brine and the rim of the jar.
- Place a lid on the top and turn it to seal the jar until it just stops turning.
- If necessary, add a sticky note to remind yourself of the ingredients, the date of making the ferments, and the pH values.
- Store the jar in a container in case of brine leak.
- Closely monitor the whole process especially in the first few days when the fermentation process is most active: the brine may leak and the ferments may float to the brine surface and be exposed to oxygen (if this happens, you will need to press down the ferments to keep them submerged).

10. Bok Choy Pickles

Time: about 10 minutes

Yields 1 pint

Pickles are more than just cucumber pickles. If a batch contains 3% salt or more, it can be considered a form of pickles. In order to try other types of pickles other than cucumber pickles, I found this recipe from *Asian Pickles* by Karen Solomon.[9]

Bok choy makes a great candidate for vegetable fermentation, as they taste crunchy and smell refreshing. The combination of garlic cloves, basil, and bok choy together make this type of ferments very flavorful and tasty. The pH of the batch of bok choy pickles I made dropped to 3.74 by Day 9. And you should be able to enjoy them in a week or less.

I would definitely recommend this to any beginner because I really liked its pleasant taste and refreshing smell when it was ready to serve.

Ingredients

- 1 pound and 1.5 ounces bok choy
- 5 teaspoons sea salt
- 5 garlic cloves
- 0.3 ounce basil
- 1 or 2 red cabbage leaves

Instructions

- Chop the bok choy into small pieces.
- Mix the basil with the bok choy and stir well.
- Add sea salt into the mix and massage for about 3-4 minutes.
- Place the garlic cloves at the bottom of jar.
- Place the mixture into the jar on the top of garlic cloves, one small batch at a time, followed by bruising and squeezing with a temper until you see that the brine is formed and covers ferments.
- Cut 1-2 small pieces of the cabbage leaves (so that they can fit into the jar) and squeeze them in on the top of ferments as the followers.
- Make sure that the ferments and the followers are submerged with 1-2 inches of space between the top of the brine and the rim of the jar.
- Place a lid on the top and turn it to seal the jar until it just stops turning.
- If necessary, add a sticky note to remind yourself of the ingredients, the date of making the ferments, and the pH values.
- Store the jar in a container in case of brine leak.
- Closely monitor the whole process especially in the first few days when the fermentation process is most active: the brine may leak and the ferments may float to the brine surface and be exposed to oxygen (if this happens, you will need to press down the ferments and the followers to keep them submerged).

11. Napa Cabbage Kimchi (Kraut Making Method) and 12. Napa Cabbage Kimchi (Presoaking Method)

Time needed: about 15 minutes (expect 8 more hours for presoaking Napa cabbage, if you are making kimchi with the presoaking method)

Yields 1 quart

As mentioned before, there are three basic ways to make kimchi: the kraut making method, the pickling method, and the presoaking method (that is, soaking cabbage overnight for about eight hours before officially making kimchi). The ingredients used for both batches here are the same. The only difference is that I presoaked cabbage for eight hours one batch before I continued with other steps. By comparison, I wanted to find out which method tasted better. Recipes are inspired by *Fermented Vegetables*.[10]

The presoaking method made more flavorful kimchi. Although made with the same ingredients, kimchi made with presoaking method tasted a bit spicier and sourer. My guess is that presoaking allows spices to penetrate into Napa cabbage, thus enhancing flavors at each bite.

I recommend you start making kimchi with the presoaking method. Either way, you can enjoy kimchi in about 5 days.

Ingredients

- 2 garlic cloves
- 2-inch ginger
- 2 pounds Napa cabbage
- 2 small red chili peppers
- ¼ jalapeno
- 1 tablespoon Korean red powder
- 2 teaspoons sea salt
- Brine (to make this brine, mix 3 teaspoons of sea salt with every cup of water; this is for presoaking method only)

Instructions

- (For presoaking method only) Add the brine into a big container and soak the Napa cabbage in that container for 8 hours.
- (For presoaking method only) Keep about 1 cup of the brine and slightly drain the soaked cabbage.
- Keep 2-3 leaves of Napa cabbage as the followers later on.
- Mince the garlic.
- Peel the ginger and cut it into small pieces.
- Cut the Napa cabbage into pieces, and mix them with minced garlic, ginger, red chili pepper, Jalapeno, and Korean red powder.
- Add 2 teaspoons of salt into the mixture and massage for 11 minutes.
- Pack everything into a jar and press them down firmly until you see the brine is formed and covers the ferments (for presoaking method, if there is not enough brine formed, you can add into the jar the preserved brine you use for presoaking).
- Cut 1-2 small pieces of red cabbage leaves, so that they can fit into the jar.
- Place them on the top of the brine, and firmly press down the leaves, so that they are submerged in the brine, too.
- Make sure that the ferments and the followers are submerged with 1-2 inches of space between the top of the brine and the rim of the jar.
- Place a lid on the top and turn it to seal the jar until it just stops turning.
- If necessary, add a sticky note to remind yourself of the ingredients, the date of making the ferments, and the pH values.
- Store the jar in a container in case of brine leak.
- Closely monitor the whole process especially in the first few days when the fermentation process is most active: the brine may leak and the ferments may float to the brine surface and be exposed to

oxygen (if this happens, you will need to press down the ferments and the followers to keep them submerged).

13. Assorted Root Kimchi

Time: about 12 minutes

Yields 1.5 quarts

Kimchi is not just Napa cabbage with spices. Any vegetables mixed with spices and 3% of salt or more can be called kimchi. Here I explored one variation of kimchi with the presoaking method. pH of ferments quickly dropped to close to 4 by Day 4. You can start enjoying the ferments in about four days. This recipe is inspired by *Wild Fermentation* by Sandor Katz,[11] and is also a beginner-friendly experiment.

Ingredients

- 10.3 ounces carrots
- 1.85 pounds daikon
- 5 red radishes
- 1 sweet onion
- 5 garlic cloves
- 2 small red chili peppers
- ½ jalapeno
- Brine (to make this brine, mix 4 cups of water with 3 tablespoons of salt)
- 2 teaspoons ginger, peeled and grinded
- 1 or 2 red cabbage leaves

Instructions

- Peel the daikon, red radishes, and carrots and chop them into small chunks.

- Soak them for 8 hours.
- Slightly drain these root vegetables.
- Peel the onion, cut it into small pieces, and set them aside.
- Mince the garlic and cut the red chili peppers and jalapeno into small pieces.
- Mix the root vegetables with all spices in a bowl.
- Pack the mixture into the jar and firmly press down the ferments until the brine is formed.
- Cut one or two small pieces of the red cabbage leaves (so that they can fit into the jar) and squeeze them in on the top of the ferments as the followers.
- Make sure that the ferments and the followers are submerged with 1-2 inches of space between the top of the brine and the rim of the jar.
- Place a lid on the top and turn it to seal the jar until it just stops turning.
- If necessary, add a sticky note to remind yourself of the ingredients, the date of making the ferments, and the pH values.
- Store the jar in a container in case of brine leak.
- Closely monitor the whole process especially in the first few days when the fermentation process is most active: the brine may leak and the ferments may float to the brine surface and be exposed to oxygen (if this happens, you will need to press down the ferments and the followers to keep them submerged).

14. Asparagus Kimchi

Time: about 6 minutes

Yields 1 quart

In this experiment, I tried the pickling method to make kimchi with the recipe inspired by *Fermented Vegetables*.[12]

SIXTEEN VEGETABLE FERMENTATION EXPERIMENTS

If you decide to start with this experiment, I recommend you add more asparagus spears into the jar until you can tightly pack them to make sure that they stand in place by themselves. I didn't add enough asparagus spears, which didn't allow them to secure their own positions inside the jar and caused them to float to the brine surface easily.

This dish can be enjoyed in about 5 days.

Ingredients

- 13 ounces asparagus, woody ends removed (about 5 inches left)
- ½ cup grated carrots
- ½ cup grated red radish
- ¼ cup grated daikon
- 1 tablespoon Korean red chili powder
- ¼ jalapeno
- 2 small red chili peppers
- 1 tablespoon grinded ginger
- 5 garlic cloves
- Cucumber brine, about 2.5 cups of cucumber juice with 3 teaspoons of sea salt per cup of juice
- 1 or 2 white cabbage leaves
- 1 teabag of black tea leaves

Instructions

- Add the carrots, red radish, daikon, red chili powder, Korean red chili powder, jalapeno, grinded ginger, garlic cloves, and black tea leaves into a blender and make the mixture into a paste.
- Arrange the asparagus spears upright in the jar.
- As you put the asparagus in, pack the paste around them as you go.
- Add the brine and make sure that all the tender tips of the spears are covered in the brine.

- Cut 1-2 small pieces of the white cabbage leaves (so that they can fit into the jar) and squeeze them in on the top of the ferments as the followers.

- Make sure that the ferments and the followers are submerged with 1-2 inches of space between the top of the brine and the rim of the jar.

- Place a lid on the top and turn it to seal the jar until it just stops turning.

- If necessary, add a sticky note to remind yourself of the ingredients, the date of making the ferments, and the pH values.

- Store the jar in a container in case of brine leak.

- Closely monitor the whole process especially in the first few days when the fermentation process is most active: the brine may leak and the ferments may float to the brine surface and be exposed to oxygen (if this happens, you will need to press down the ferments and the followers to keep them submerged).

15. Fermented Red Cabbage (with a Regular Lid) and 16. Fermented Red Cabbage (with an Airlock)

At last, in order to learn more about whether the use of different lids would influence the taste or texture of ferments, I designed these two experiments. I used a metal lid for one bottle of ferments and an airlock for another, while following the kraut making method with the use of 3% salt. And, I found no difference in flavors or textures between two bottles.

These two experiments helped me learn more about red cabbage fermentation: after mixing salt with the sliced red cabbage leaves and massaging them for ten minutes or even longer, I only started to see very little red cabbage juice formed; it took me eight minutes to massage one pound of the sliced white cabbage leaves to form more-than-enough juice for the brine. Therefore, be prepared that it takes longer for the red cabbage leaves to form the brine. It is not impossible to create the brine; you just need more patience.

Because of this, less brine will be produced for the red cabbage fermentation (if you don't introduce more liquid from the outside) and you may even see the brine is "evaporated" after you transport ferments to the fridge and let them sit for a few days). I reached out to Kirsten Shockey, author of *Fermented Vegetables*, and asked her whether ferments were still edible that way. The answer is yes. She recommends you press down the veggies tightly and keep the ferments in jars that don't have a lot of air space on the top of the ferments if you are not eating them very quickly.

Additionally, this is another good example to show that fermented vegetables indeed taste better after staying in the fridge for some time: fermented red cabbage develops better flavors and becomes crunchier after it sits in the fridge for about a week or longer. And I actually preferred the taste of these ferments after they sat in the fridge for a while instead of consuming them immediately after the pH reached 4.6 on the kitchen counter.

Although they may not be my most recommended getting-started recipes for you (because the preparation is a bit time-consuming), they are definitely worth a try down the road. The pH of the fermented red cabbage I made dropped to 4.68 by Day 3, and to 3.64 by Day 9. If you follow the instructions below, you should be able to enjoy them in around five days.

The following recipe can be used to both experiments.

Time: about 17 minutes

Yields ½ pint

Ingredients

- 8 ounces red cabbage, remove outer leaf if cabbage is not organic
- 1.5 teaspoons sea salt

Instructions

- Discard the outer leaf if the cabbage is not organic and keep 1-2 whole leaves as the followers later on.
- Chop the rest of red cabbage into slices.
- Put the red cabbage leaves into a bowl, mix in sea salt, and massage them for 10 minutes, until you see the brine is slowly formed.

- Pack the vegetables into the jar with a temper (or, its alternative), continue to bruise the leaves to let more juice come out, and press down the cabbage firmly.
- When enough brine is produced to cover the vegetables, place the small pieces of cabbage leaves on the top and squeeze them in on the top of the ferments as the followers.
- Make sure that the ferments and the followers are submerged with 1-2 inches of space between the top of the brine and the rim of the jar.
- Place a lid on the top and turn it to seal the jar until it just stops turning.
- If necessary, add a sticky note to remind yourself of the ingredients, the date of making the ferments, and the pH values.
- Store the jar in a container in case of brine leak.
- Closely monitor the whole process especially in the first few days when the fermentation process is most active: the brine may leak and the ferments may float to the brine surface and be exposed to oxygen (if this happens, you will need to press down the ferments and the followers to keep them submerged).

Chapter 13: Incorporating Fermented Vegetables into Daily Life

"After I incorporated these foods into my life they began to change me from the inside out, and now I cannot go back."

– DONNA SCHWENK
Cultured Food for Health

Home-Made Probiotic Dishes at a Glance

1. Spicy Probiotic Beet Juice
2. Sauerkraut Sandwich
3. Vegetable Kimchi Fried Rice
4. Nutty Quinoa Breakfast
5. Kraut Avocado Dip
6. Coconut Chicken and Veggie Stew
7. Vietnamese Spring Rolls with Tofu and Cucumber Pickles
8. Probiotic Tzatziki
9. Coconut Delight
10. The Three-Minute Egg Breakfast

1. Spicy Probiotic Beet Juice

Time: about 5 minutes

Serves 1

Probiotic source: beet kvass (find "Beet Kvass" from Chapter 12 for the recipe.)

The sweetness from the apple, cucumber, and celery stalks, the sourness from the beet kvass, and the spicy flavor from the ginger mix well in this juice. I recommend you drink this early in the morning before you have breakfast to maximize the absorption of the nutrients. Personally, I also treat this as an appetizer as it will help with the digestion of whatever foods you eat later on.

Ingredients

- 1 long cucumber, peeled if not organic
- 2 celery stalks
- 4-inch ginger, peeled
- 1 small gala apple, peeled if not organic

- 2 tablespoons beet kvass

Instruction

- Put everything except beet kvass, piece by piece, into the juicer.
- Add beet kvass into the juice and mix well.

2. Sauerkraut Sandwich

Time: about 3 minutes

Serves 3

Probiotic source: white cabbage sauerkraut or red cabbage sauerkraut (find "Salted White Cabbage Sauerkraut" and "Fermented Red Cabbage" from Chapter 12 for the recipes.)

This breakfast has a good balance of probiotics (from the krauts), dietary fiber (from the cucumber and tomato), and protein (from the whole grain bread and peanut butter). These days, I have been mindful about eating probiotics in combination with foods rich in dietary fiber to maximize the digestion as well as feed the microbes inside my body. This gives you an example of how you can introduce both fermented foods and dietary fiber into the body at the same time.

Ingredients

- ¼ of a long cucumber, peeled and cut into small pieces
- ½ tomato, cut into small pieces
- 3 pieces whole grain bread
- 1 teaspoon goat cheese for each piece of bread
- 2 teaspoons peanut butter for each piece of bread
- 1 tablespoon sauerkraut of your choice for each piece of bread

Instructions

- Slightly toast the whole grain bread.
- While toasting the bread, mix the cucumber, tomato, and sauerkraut together in a bowl, and mix well.
- Spread an equal amount of peanut butter on the top of each piece of the bread.
- Evenly distribute the mixed ingredients in the bowl onto the bread.
- Divide the goat cheese into tiny pieces and sprinkle them on the top of mixed ingredients.

3. Vegetable Kimchi Fried Rice

Time: about 40 minutes (plus 8 more hours for presoaking brown rice and cashews overnight)

Serves 2

Probiotic source: fermented asparagus (find "Asparagus Kimchi" from Chapter 12 for the recipe.)

There are some extra steps you need to do one day before you heat up your pan and start cooking. These additional procedures are important because they help you maximize the nutrient intake from grains and nuts.

To cook the rice, I recommend presoaking it overnight. Similarly, it's better if you could presoak the nuts overnight before use. I learned these tricks from *Nourishing Traditions*.[1] Presoaking grains and nuts helps remove phytic acid for better digestion and absorptions of minerals and other nutrients.

Additionally, I recommend using avocado oil for high temperature cooking; and, whenever salt is needed, I always use Himalayan sea salt because of its rich mineral content.

Ingredients

- ½ cup brown rice

- 1 cup warm filtered water mixed with 1 tablespoon yogurt
- ¼ cup cashews
- 1 cup warm salted filtered water
- 1 green bell pepper (with seeds removed and the pepper chopped into small pieces)
- 1 cup chopped carrots
- 3 garlic cloves (minced)
- 1 tablespoon finely chopped ginger root
- 2 cups peeled and diced daikon
- 2 eggs (from pastured chicken, if possible)
- 1 medium sized onion (peeled and chopped)
- 2 tablespoons avocado oil
- ½ cup filtered water
- 6 fermented asparagus spears
- Himalayan sea salt, to taste
- Ground black pepper, to taste

Instructions

- Soak the brown rice in warm filtered water mixed with yogurt overnight (or, 8 hours); and soak the cashews in warm salted filtered water in a separate container for the same amount of time.
- Drain the brown rice and cashews, rinse them under filtered water, drain them again, and set the nuts aside.
- Add ½ cup of filtered water and soaked brown rice into a pot, bring the water to boil, and simmer for 30 minutes (or, until rice becomes soft); remove the rice from heat.
- After rice is simmered for 20 minutes, heat up a frying pan with medium to high heat, add avocado oil, minced garlic, chopped

ginger root, and chopped onion into the pan, and sauté until you can smell the aroma from the onion.

- Add the diced daikon and chopped carrots into the pan and sauté for 3-4 minutes (or, when daikon and carrots become soft) and add more filtered water if the pan is too dry.

- Add the prepared bell pepper and soaked cashews into the mixed veggies and continue to sauté for 3-4 minutes.

- Mix in the cooked brown rice, sauté until all ingredients are blended together.

- Crack two eggs into the fried rice and continue to sauté for about 1 minute.

- Add pinches of salt and pepper and mix well.

- Transport the fried rice to two plates and, when the rice cools down, place 3 asparagus spears on the top of each plate.

4. Nutty Quinoa Breakfast

Time: about 25 minutes

Serves 2-3

Probiotic source: white cabbage sauerkraut or red cabbage sauerkraut (find "Salted White Cabbage Sauerkraut", and "Fermented Red Cabbage" from Chapter 12 for the recipes.)

What I like the most about this breakfast is that the chewy texture of the quinoa goes very well with the crunchiness of the krauts. The key to make nutty quinoa is adding the right amount of water and cooking this pseudo-grain for the right amount of time. If you want a softer texture, add more water and let quinoa sit for a longer period of time after you turn off heat but before you lift the lid.

Ingredients

- 1 cup quinoa
- 2 cups warm filtered water mixed with 1 tablespoon yogurt
- ¼ cup pine nuts
- ½ cup salted warm filtered water
- 1 cup filtered water
- 2 tablespoons olive oil
- 1 few pinches Himalayan salt
- A few pinches black pepper
- ½ cup kraut of your choice

Instructions

- Soak quinoa in warm filtered water mixed with yogurt overnight (or, about 8 hours) and soak the pine nuts in salted warm filtered water in a separate container for the same amount of time.
- Drain the quinoa and nuts separately, rinse them under filtered water, drain them well, and set the nuts aside.
- Mix the 1 cup of filtered water with the soaked quinoa in a pot, and bring it to boil.
- Simmer for 8 minutes.
- Turn off the heat and let it sit for 3-4 minutes.
- Lift the lid and stir well to help the heat escape and to keep the quinoa from being stuck to the bottom of the pot.
- Let it sit without the lid until quinoa cools.
- Add the olive oil, salt, black pepper, and pine nuts, and mix well.
- Add the kraut of your choice and mix well.

5. Kraut Avocado Dip

Time: about 2 minutes

Yields about 1 cup

Probiotic source: white cabbage sauerkraut or red cabbage sauerkraut (find "Salted White Cabbage Sauerkraut", and "Fermented Red Cabbage" from Chapter 12 for the recipes.)

Our family love having sesame seed oil as it can greatly enhance the flavor of any dish. If you haven't tried guacamole with sesame seed oil, then you should definitely give it a try. What I also like about this dip is that the soft texture of avocado paste and the crunchiness of the krauts make a good combo.

Ingredients

- 2 avocados, peeled, pitted, and cut into small chunks
- 3 cloves garlic (minced)
- ½ tomato (diced)
- Juice from ½ lemon
- 2 tablespoons sauerkraut
- ½ teaspoon sesame seed oil
- ½ small onion, peeled and chopped
- ¼ teaspoon Himalayan pink salt

Instructions

- Put the avocado into a blender and blend it into a paste.
- Scoop the paste into a bowl and add the garlic, tomato, lemon juice, kraut of your choice, sesame oil, onion, and salt.

6. Coconut Chicken and Veggie Stew

Time: about 30 minutes

Serves 3

Probiotic source: kimchi (find the pre-soaking method of making "Napa Cabbage Kimchi" from Chapter 12 for the recipe.)

What I like about this dish is that the spicy flavors from kimchi mix very well with the coconut milk. If you would like a creamier texture and a stronger coconut flavor in the stew, I would recommend you do not dilute the milk with an extra cup of water. You can also try doubling the amount of kimchi, if your body can tolerate fermented vegetables very well. This dish will go with brown rice or spaghetti.

Ingredients

- 2 tablespoons avocado oil
- 2 tablespoons chopped ginger root
- ½ cup chopped onion
- 4 cups sliced bell peppers
- 1 ½ cups chopped leek
- 2 cups chopped potatoes (potatoes are chopped into sizes of ¼ of a red radish; keep the skin if they are organic)
- 1 ½ cup organic chicken breast (cut into strips)
- 1 tablespoon lemon juice
- 2 cups filtered water
- 2 cups organic coconut milk
- Himalayan salt, to taste
- 3 tablespoons kimchi

Instructions

- Add the chopped potatoes and 1 cup of water into a small pot, bring the water to boil, and simmer for 10 minutes (or, until potatoes become soft.
- After simmering the potatoes for about 5 minutes, add avocado oil into a large pan and heat up the oil with medium to high heat.

- Add the chopped ginger and onion into the pan and cook until you can smell the onion.

- Add chicken breast into the pan and sauté until you see the pink color turns solid white.

- Mix in the bell peppers and chopped leek and sauté with medium heat for about 1 minute.

- Mix organic coconut milk with 1 cup of filtered water, pour the diluted milk into the large pan, and bring the coconut milk to boil.

- Lower the heat, add lemon juice, and simmer for 15 minutes (mix in the potatoes when they are cooked and become soft).

- When there are a few minutes left, add pinches of salt, mix well, and continue to simmer for another few minutes.

- Turn off heat and, when the stew cools down, add kimchi and mix well.

7. Vietnamese Spring Rolls with Tofu and Cucumber Pickles

Time: about 30 minutes

Yields 8 spring rolls

Probiotic source: cucumber pickles (find "Cucumber Pickles (with Tannins)" from Chapter 12 for the recipe.)

I love Vietnamese spring rolls! They are yummy and easy to make. Here is the catch: you should have a nice control of the strength when you fold the rice papers, because they are soft and delicate once dipped into the warm water. If you have never made spring rolls before, I actually recommend you use two rice papers as one wrapper to make one spring roll. If that is what you plan to do, expect twice the amount of the time for softening one wrapper. You can serve the spring rolls with diluted fish sauce or peanut butter.

Ingredients

- 8 pieces of rice paper (inch in diameter)
- 2 cups filtered water
- 2 large Romaine lettuce leaves, sliced
- Firm tofu, enough to make 8 3.5-inch strips (on either side of the tofu strip is a 0.5-inch-by-0.5-inch square)
- 1 ounce rice noodles (about ½ cup after cooked and drained)
- 16 basil leaves
- 3.5-inch of one cucumber pickle, cut into strips
- 2 tablespoons avocado oil

Instructions

- Bring the water to boil; boil the rice noodles for about 5 minutes, drain, and set them aside.
- Add the oil to a small frying pan and heat up the oil with medium to high heat.
- Add the tofu strips into the frying pan and cook each side evenly until they turn golden.
- Turn off the heat and put the fried tofu aside.
- Fill a large bowl with warm water; carefully dip one rice paper into the water for a few seconds until the paper starts to become soft; lay the rice paper flat on a clean kitchen towel.
- On the rice paper about 1.5 inches above the center, place 1 tofu strip, a handful of drained rice noodles, a handful of sliced lettuce leaves, a few cucumber pickles strips, and 2 basil leaves.
- Fold the left and the right uncovered parts towards the center; fold the top uncovered part over the filling; tightly roll the paper from the top to the bottom; repeat the rest with remaining ingredients.

8. Probiotic Tzatziki

Time: about 2 minutes

Yields about ¾ cup

Probiotic source: yogurt and dill cucumber brine (find "Cucumber Pickles (with Tannins)" from Chapter 12 for the recipe.)

My inspiration comes from the traditional Greek dip called tzatziki, a healthy yogurt-based sauce used to serve meat. I first tried this at a Greek restaurant, immediately fell in love with it, and wanted to be able to make it at home. One day, I did a quick search online and learned that a traditional tzatziki contains cucumbers and dill, I immediately thought of the dill cucumber pickles that were sitting inside the fridge. Then this dish was born. It has the flavor and texture of a traditionally-made tzatziki. The different is that this dip has additional friendly bacteria that are good to your gut! This dip goes very well with fried fish, cooked red meat, and grilled chicken.

Ingredients

- ½ cup full fat plain Greek yogurt
- 1 teaspoon olive oil
- 1 tablespoon garlic dill pickle brine
- Black pepper, to taste
- Himalayan salt, to taste
- ½ medium cucumber, chopped
- 2 teaspoons lemon juice
- 3 cloves of garlic, minced (if you prefer less strong garlic flavors, use only 1 clove)
- Cilantro (optional)
- Pickles, chopped to small pieces (optional)

Instruction

- Mix everything into a bowl and serve with seafood or any meat of your choice.

9. Coconut Kraut Delight

Time: about 30 seconds

Serves 1

Probiotic source: red cabbage sauerkraut (find "Fermented Red Cabbage" from Chapter 12 for the recipe).

This is a very simple snacking idea that takes you only a few seconds to make. My takeaway here is that: sometimes you don't need fancy procedures to give your taste buds a pleasant surprise. Personally, I really like the flavors and textures of the red cabbage kraut and the coconut flakes combined.

Ingredients

- 1 tablespoon red cabbage kraut
- 2 tablespoons coconut flakes

Instruction

- Mix them together and enjoy right away; alternatively, you serve this Delight with crackers.

10. The Seven-Minute Egg Breakfast

Time: about 13 minutes

Serves 1

Probiotic source: eggplant garlic kraut (find "Eggplant Garlic Kraut" from Chapter 12 for the recipe.)

Taking a close look at the ingredients of this dish, you will find a well-mixed of nutrients including protein (from the egg), dietary fiber (from mixed greens), good fats (from olive oil), and probiotics (from the eggplant garlic kraut). The egg yolk makes a really good salad dressing for the greens and the kraut. To enhance your eating experience, you can serve this dish with mindfulness and gratitude for yourself taking the time to prepare such a wholesome meal for the body.

Ingredients

- 1 egg (ideally, organic)
- 1 to 2 cups salad with mixed greens
- 1 tablespoon eggplant garlic kraut
- 2 cups filtered water
- 1 tablespoon olive oil
- Salt, to taste
- Pepper, to taste

Instructions

- Bring the water to boil (in the meantime, place the mixed greens in a bowl and mix them with the eggplant garlic kraut).
- Turn off heat, put the egg into the pot, cover it with a lid, and continue to cook for 7 minutes.
- Immediately take the egg out of the water and rinse it under filtered water for a few seconds.
- Once the egg cools down, peel off the egg shell, and place it on the top of the mixed greens.
- Cut the egg into halves, so that runny egg yolk comes out.
- Drizzle the olive oil on the top of the egg and mixed greens.

- Add pinches of salt and pepper (you may consider skipping this, as the kraut is already salty).

METRIC CONVERSION TABLES

Approximate Equivalents by Weight		
Ounce (oz)	Pound (lb)	Gram (g)
1	1/16	30
4	¼	120
8	½	240
12	¾	360
16	1	480

Approximate Equivalents by Length	
Inch (in)	Centimeter (cm)
1	2.5
2	5
3	7.5
4	10
5	12.5

Approximate Equivalent by Volume					
Teaspoon (tsp)	Tablespoon (tbsp)	Fluid Ounce (fl oz)	Cup	Pint	Quart
1	1/3	1/6	-	-	-
3	1	1/2	-	-	-
6	2	1	1/8	-	-
-	16	8	1	½	¼
-	-	16	2	1	1/2
-	-	32	4	2	1

Temperature Equivalents	
Celsius (°C)	Fahrenheit (°F)
0	32
10	50
15	59
20	68
25	77
30	86
35	95
40	104
45	113
50	122

RESOURCES

The following resources are grouped into five categories, which you can browse based on your interests. All resources can also be found on my website at www.tracyhuang.me/vegetable-fermentation, which will be updated as I learn more along the way.

Specifically, these five categories are: 1) **"Foods"**, to help you learn more about real foods and food movements; 2) **"Human and Microbes"**, to help you understand your body and microbes and to discuss where the study of microbiology is leading us to; 3) **"Food Science on Fermentation"**, to explore science behind fermentation and microbial activities in foods; 4) **"Fermented Foods"**, to discover healing benefits of fermented foods, to master techniques taught by fermentation experts, and to check out creative established recipes; and 5) **"Tools"**, to look at products I recommend to help you get started with your fermentation experiments.

Foods

[Book] *Food Rules* by Michael Pollan
This is a book that inspires us to look back to what our ancestors ate, respect our eating traditions, and use the traditional food rules as a reference for healthy eating. Just as Michael Pollan points out, it is totally okay to still eat healthy without knowing what antioxidants are.

[Book] *Nourishing Traditions: The Cookbook that Challenges Politically Correct Nutrition and the Diet Dictocrats* by Sally Fallon and Mary G. Enig
A lot of people say this book has changed their perceptions about foods. I grew up eating a lot of what the book recommends – such as fish soups, bone broths, and lard – in China, and did not continue doing so after I came to the US to study. This book enhances my belief that we should connect ancient wisdom and modern food science to achieve our optimal health and wellbeing. The book also touches on a lot of trendy topics – such as alkaline diets, veganism, and eating according to your blood type – and gives a fair point of view on how we should look at them.

[Documentary] *Food Inc.* directed by Robert Kenner
This is an eye-opening film that reveals what goes behind the foods you are eating; it delivers the messages you should be mindful of what you eat and offers helpful guides on how to be proactive about taking control of your own health.

[Organization] Weston A Price
This is an organization dedicated to food research and education. Sally Fallon, co-author of *Nourishing Traditions*, is the president of this organization. You can visit the organization at www.westonaprice.org.

[Organization] Environmental Working Group
This site gives you a lot of valuable education on how you should shop, what to look for, and what to watch out for as you buy foods. You can visit the organization at www.ewg.org.

Human and Microbes

[Video] "The Invisible Universe of the Human Microbiome"
This is a five-minutes-and-28-seconds animated video that talks about what microbes are, what they do, and why they are important to you. You can watch the video at http://bit.ly/invisible-universe.

[Video] "How Our Microbes Make Us Who We Are" by Rob Knight
With a lot of compelling visual demonstrations, this microbial ecologist gives you a 17-minute-and-24-second talk to help you understand the microbes inside you. You can watch the video at http://bit.ly/human-microbes.

[Book] *The Good Gut: Take Control of Your Weight, Your Mood, and Your Long Term Health* by Justin Sonnenburg and Erica Sonnenburg
This is a great book that gives you an overview of what is going on right now in the modern microbial studies. You will have a big picture and clearly see why fermented foods have become such a hot topic these days. The book covers what microbes are, why they are important, how they are co-existing with us, what we need to do to better co-exist with them, what to eat in daily life, and where the study of microbiome is leading us to.

Food Science on Fermentation

[Website] Pickle Bibliography
This is a very useful website that documents all the research papers published by food scientist and microbiologist Fred Breidt and his team. Topics are related to vegetable fermentation. If you are concerned about making and eating fermented vegetables, this is a great website to subscribe to and visit regularly. Visit http://bit.ly/pickle-bibliography for more details.

[Website] Microbial Foods

This is another interesting site that explores the science behind why fermentation happens in foods, the biology of important microbial groups in different fermented foods, and related trendy topics. Learn more at www.microbialfoods.org.

Fermented Foods

[Book] *The Art of Fermentation* by Sandor Katz

This is a cookbook but more than just a cookbook; it is also a very comprehensive introduction to how fermentation connects elements like people, foods, plants, animals, society, art, culture, farming and more. If you want to understand the history of fermentation and how its role has been played in the human history and the current society, then reading this book allows you to start your journey with a 3,000-foot view on these topics.

[Book] *Wild Fermentation* by Sandor Katz

This is a book that tells you not to be intimidated by fermentation but to try out easy and simple techniques shared in this book. Katz also briefly talks about how fermentation has taught him to learn about life and death. Thanks to this book and *The Art of Fermentation*, I have developed a deeper understanding of fermentation: this ancient practice not only nourishes the body physically, but also helps one reflects on life.

[Book] *Fermented Vegetables* by Kirsten Shockey and Christopher Shockey

This is one of my favorite books on fermentation because of the included clear instructions. This book also teaches me to go beyond science – that is, the nutritional value of fermented vegetables – and to connect with my intuition by creating my own recipes that will allow me to fully appreciate the flavors and colors of the vegetables.

[Book] *Cultured Foods for Life* by Donna Schwenk

By specifically talking about the "trilogy" – kefir, cultured vegetables, and kombucha – the author shares with you how these fermented foods have changed her life and can change yours, too. I first heard of kefir from this book; and now I am making my own kefir, drinking it frequently, and enjoying better digestion and better skin.

[Book] *Cultured Foods for Health* by Donna Schwenk

Following the messages from her last book, Schwenk wrote this book with two new key messages: 1) consuming assorted fermented foods can heal the body in many ways; and 2) it is crucial to eat fermented foods with prebiotics (the food that feeds the microbes inside you) such as dietary fiber-rich foods.

[Book] *Real Food Fermentation* by Alex Lewin

Lewin makes fermentation fun and easy with clear instructions and a lot of visual presentations to teach you how to make different kinds of fermented foods step by step.

[Audio] Hay House Radio Tuesday Talks by Donna Schwenk

Schwenk talks about the healing benefits of the fermented foods in this radio station every Tuesday afternoon from 1 to 2 EST. At the end of the talk, you can even dial in to ask her questions, which is a great way to get your questions answered by experts. Listen to the radio at http://bit.ly/hay-house-radio-tuesday.

[Organization] Boston Ferments

Hosting a series of workshops, meetups, and an annual festival, this is a Boston-based non-profit organization dedicated to helping people understand fermentation and participate in related activities. Visit www.bostonferments.com for more.

[Event] Boston Fermentation Festival

This is an annual festival hosted by Boston Ferments. If you living in or around Greater Boston Area, I highly recommend you check out this festival, where you participate in lectures by experts and attend hands-on workshop. The goal is to allow you to connect with like-minded people, have fun with making and eating assorted fermented foods, and get educated on related topics. If you cannot conveniently get access to this festival, try searching around online to see if you can join any local community with similar activates.

[Website] Cultures for Health

This is an online store where you learn how to make different fermented foods and how to advance your fermentation practice. I first visited this site to buy kefir grains and starter cultures for making kombucha and sourdough. Later, I became a frequent visitor as I love reading the helpful articles; I have also received a lot of non-product-related professional support from the customer service representatives when I need trouble-shooting advice. You can visit the store at www.culturesforhealth.com.

Tools

You may visit www.tracyhuang.me/vegetable-fermentation for details.

Knife
Tramontina Proline Santoku Knives

Cutting Board
KitchenAid Non-Slip Badge Logo Cutting Board, 8x10-inch

Peeler
OXO Good Grips Swivel Peeler

Ziploc Bags
Ziploc Sandwich Bags, Pack of 150, 6.5x5.867-inch

One-Quart Mason Jars
Ball Regular Mouth Quart Jars with Lids and Bands, Set of 12

One-Pint Mason Jars
Ball Jar 1pt Mason Jars, Case of 12

Half-a-Pint Mason Jars
Ball Regular Mouth Half Pint Jars with Lids and Bands, Set of 12

pH Test Paper, 3.0 – 5.5pHs
Micro Essential Lab Hydrion Short Range pH Test Paper Dispenser, 3.0-5.5 pH

pH Test Meter
Etekcity High Accuracy Pocket Size Handheld pH Meter Pen Tester

pH Buffer Capsules, 8.00
Hydrion Buffer Capsule 8.00

pH Buffer Capsules, 4.00
Hydrion Buffer Capsule 4.00

Food Processor
KitchenAid 3-Speed Hand Blender – Contour Silver

Garlic Mincer
Kuhn Rikon Epicurean Garlic Press

Airlocks
FARMcurious Mason Jar Fermenting Kit Mold Free, 2 Pack

END NOTES

Chapter 1

1. Justin Sonnenburg and Erica Sonnenburg, "The Microbial World," chap. 1 in *The Good Gut* (New York: Penguin Press, 2015), Google Play Books.
2. Rob Knight, "How Our Microbes Make Us Who We Are," filmed February 2014, TED video, 17:24, www.ted.com/talks/rob_knight_how_our_microbes_make_us_who_we_are.
3. Michael Pollan, "Some of My Best Friends Are Germs," *New York Times*, May 15, 2013, www.nytimes.com/2013/05/19/magazine/say-hello-to-the-100-trillion-bacteria-that-make-up-your-microbiome.html?_r=4.

4. S. E. Gould, "How Bacteria Break Down Human Food," *Scientific American*, June 24, 2012, blogs.scientificamerican.com/lab-rat/how-bacteria-break-down-human-food/.

5. David Williams, "Probiotic Species and Strains: What Are Their Difference?," *DrDavidWilliams.com*, September 9, 2015, www.drdavidwilliams.com/probiotic-strains/.

6. University of Michigan Health System, "Investigating the Fiber of Our Being: How Our Gut Bacteria Metabolize Complex Carbohydrates from Fruits, Vegetables," *ScienceDaily*, February 12, 2014, www.sciencedaily.com/releases/2014/02/140212132851.htm.

7. Bob Stein, "Gut Bacteria Might Guide the Workings of Our Minds," *New Hampshire Public Radio*, November 18, 2013, www.npr.org/sections/health-shots/2013/11/18/244526773/gut-bacteria-might-guide-the-workings-of-our-minds.

8. Mark Lyte, "Microbial Endocrinology in the Microbiome-Gut-Brain Axis: How Bacterial Production and Utilization of Neurochemicals Influence Behavior," ed. Virginia Miller, *PLoS Pathog* 9, no. 11 (November 14, 2013): e1003726, doi: 10.1371/journal.ppat.1003726.

9. Ibid.

10. Sandra Blakeslee, "Complex and Hidden Brain in Gut Makes Stomachaches and Butterflies," *New York Times*, January 23, 1996, www.nytimes.com/1996/01/23/science/complex-and-hidden-brain-in-gut-makes-stomachaches-and-butterflies.html.

11. Charles L. Raison, Christopher A. Lowry, and Graham A. W. Rook, "Inflammation, Sanitation, and Consternation: Loss of Contact with Coevolved, Tolerogenic Microorganisms and the Pathophysiology and Treatment of Major Depression," abstract, *Archives of General Psychiatry* 67, no. 12 (December 2010): 1211-24, doi: 10.1001/archgenpsychiatry.2010.161.

12. John B. Furness, Wolfgang A. A. Kunze, and Nadine Clerc, "Nutrient Tasting and Signaling Mechanisms in the Gut. II. The Intestine as a Sensory Organ: Neural, Endocrine, and Immune Responses," abstract, *American Journal of Physiology* 277, no. 5 (November 1, 1999): G922-G928, www.ncbi.nlm.nih.gov/pubmed/10564096.

13. G. Vighi, F. Marcucci, L. Sensi, G. Di Cara, and F. Frati, "Allergy and the Gastrointestinal System," supplement, *Clinical and Experimental Immunology* 153, no. S1 (September 2008): 3-6, doi: 10.1111/j.1365-2249.2008.03713.x.
14. Albert-Ludwigs-Universität Freiburg, "Natural Intestinal Flora Strengthen Immune System," *ScienceDaily*, July 2, 2012, www.sciencedaily.com/releases/2012/07/120702152940.htm.
15. Joseph Mercola, "The Healing Power of Probiotics Impresses Researchers," *Mercola.com*, October 11, 2010, articles.mercola .com/sites/articles/archive/2010/10/11/probiotics-healing-power-impresses-researchers.aspx.
16. Stephen Daniells, "Breakthrough Study Shows Personalised Nutrition Future for Probiotics," *NUTRAIngredients.com*, September 15, 2010, www.nutraingredients.com/research/ breakthrough-study-shows-personalised-nutrition-future-for-probiotics.
17. Mercola, "Healing Power of Probiotics."
18. F. Purchiaroni, A. Tortora, M. Gabrielli, F. Bertucci, G. Gigante, G. Ianiro, V. Ojetti, E. Scarpellini, and A. Gasbarrini, "The Role of Intestinal Microbiotia and the Immune System," abstract, *European Review for Medical and Pharmacological Sciences* 17, no. 3 (February 2013): 323-33, www.ncbi.nlm.nih. gov/pubmed/23426535.
19. Catharine Paddock, "Gut Bacteria Essential for Immune Cell Development," *Medical News Today*, March 13, 2014, www. medicalnewstoday.com/articles/273958.php.
20. Jeffrey Gordon, interview by Ira Flatow, *New Hampshire Public Radio*, podcast audio, September 6, 2013, www.npr.org/2013/ 09/06/219669536/do-your-gut-bacteria-influence-your-metabolism.
21. Claudia Wallis, "How Gut Bacteria Help Make Us Fat and Thin," *Scientific American*, June 1, 2014, www.scientificamerican. com/article/how-gut-bacteria-help-make-us-fat-and-thin/.
22. Klara Sjögren, Cecilia Engdahl, Petra Henning, Ulf H. Lerner, Valentina Tremaroli, Marie K. Lagerquist, Fredrik Bäckhed, and Claes Ohlsson, " The Gut Microbiota Regulates Bone

Mass in Mice," *Journal of Bone and Mineral Research* 27, no. 6 (June 2012): 1357-67, doi: 10.1002/jbmr.1588.

23. Ibid.

24. Fredrik H. Karlsson, Frida Fåk, Intawat Nookaew, Valentina Tremaroli, Björn Fagerberg, Dina Petranovic, Fredrik Backhed, and Jens Nielsen, "Symptomatic Atherosclerosis Is Associated with an Altered Gut Metagenome," *Nature Communications* 3 (December 4, 2012): 1245, doi: 10.1038/ncomms2266.

25. Joseph Mercola, "You Need Vitamin K to Prevent Arterial Plague & Heart Disease," *Mercola.com*, May 14, 2003, articles.mercola.com/sites/articles/archive/2003/05/14/vitamin-k-part-one.aspx.

26. Blakeslee, "Complex and Hidden Brain."

27. Ronald S. Smith, "The Immune-Brain Connection", chap. 1 in *Cytokines and Depression* (n.p., 2010), www.cytokines-and-depression.com/chapter3.html.

28. Blakeslee, "Complex and Hidden Brain."

29. Joseph Mercola, "If You Can't Beat Depression, Your Gut Bacteria Could be the Reason," *Mercola.com*, April 12, 2011, articles.mercola.com/sites/articles/archive/2011/04/12/beware--bacteria-growing-in-your-gut-can-influence-your-behavior.aspx.

30. "Gut-Brain Interrelationships and Control of Feeding Behavior," *The Medical Biochemistry Page*, accessed June 13, 2016, www.themedicalbiochemistrypage.org/gut-brain.php.

31. Rachel Champeau, "Changing Gut Bacteria through Diet Affects Brain Function, UCLA Study Shows," *UCLA News Room*, May 28, 2016, newsroom.ucla.edu/releases/changing-gut-bacteria-through-245617.

32. Adam Hadhazy, "Think Twice: How the Gut's 'Second Brain' Influences Mood and Well-Being," *Scientific American*, February 12, 2010, www.scientificamerican.com/article/gut-second-brain/.

33. Partnership for Environmental Education and Rural Health, "Digestive System: Why It Matters", *Texas A&M University*,

accessed June 14, 2016, peer.tamu.edu/curriculum_modules/OrganSystems/module_2/whyitmatters.htm.

34. Moises Velasquez-Manoff, "Are Happy Gut Bacteria Key to Weight Loss?," *Mother Jones*, April 22, 2013, www.motherjones.com/environment/2013/04/gut-microbiome-bacteria-weight-loss.

35. North Shore-Long Island Jewish Health System, "How the Immune System and Brain Communicate to Control Disease," *ScienceDaily*, accessed June 14, 2016, www.sciencedaily.com/releases/2008/07/080721173748.htm.

36. W. Bowe, N. B. Patel, and A. C. Logan, "Acne Vulgaris, Probiotics and the Gut-Brain-Skin Axis: from Anecdote to Translational Medicine," abstract, *Beneficial Microbes* 5, no. 2 (July 1, 2014): 185-99, doi: 10.3920/BM2012.0060.

37. "Our Practitioners," *Ancient Path Acupuncture and Herbs*, accessed June 14, 2016, www.ancientpathweb.com/practitioners.

38. Michael Pollan, introduction to *Food Rules* (New York: Penguin Books, 2009).

39. Weston A. Price Foundation, accessed June 14, 2016, http://www.westonaprice.org/.

40. Sally Fallon and Mary G. Enig, preface to *Nourishing Traditions: The Cookbook that Challenges Politically Correct Nutrition and the Diet Dictocrats*, 2nd ed. (Washington, DC: NewTrends Publishing, 1999).

Chapter 2

1. Sufang Guo, Sabu S. Padmadas, Fengmin Zhao, James J. Brown, and R. William Stones, "Delivery Settings and Caesarean Section Rates in China," *World Health Organization* 85, no. 10 (October 2007): 733-820, doi: 10.2471/BLT.06.035808.

2. Josef Neu and Jona Rushing, "Cesarean Versus Vaginal Delivery: Long Term Infant Outcomes and the Hygiene Hypothesis,"

      Clinics in Perinatology 38, no. 2 (June 2011): 321-31, www.ncbi.nlm.nih.gov/pmc/articles/PMC3110651/.
3. Ibid.
4. "Boost C-section Babies by Giving Them Vaginal Bacteria," *New Scientist*, February 1, 2016, www.newscientist.com/article/2075768-boost-c-section-babies-by-giving-them-vaginal-bacteria/.
5. Sonnenburg and Sonnenburg, "Our First Microbiota Inhabitants," chap. 2 in *Good Gut* (see chap. 1, n. 1).
6. Ibid.
7. Ethan A. Huff, "Nine Reasons to Never Eat Processed Foods Again," *Natural News*, April 2, 2013, www.naturalnews.com/039743_processed_foods_eating_reasons.html.
8. Joseph Mercola, "9 Ways That Eating Processed Food Made the World Sick and Fat," *Mercola.com*, February 12, 2014, articles.mercola.com/sites/articles/archive/2014/02/12/9-dangers-processed-foods.aspx.
9. Mark Hyman, "Is High Fructose Corn Syrup Really that Bad for You?," *Dr.Hyman.com*, October 24, 2014, www.drhyman.com/blog/2014/10/24/manufacturers-downsized-high-fructose-corn-syrup-still-concerned/.
10. Sonnenburg and Sonnenburg, "Microbes as an Extension of the Mucosal Immune System," chap. 3 in *Good Gut* (see chap. 1, n. 1).
11. Débora Estadella, Claudia M. da Penha Oller do Nascimento, Lila M. Oyama, Eliane B. Ribeiro, Ana R. Dâmaso, and Aline de Piano, "Lipotoxicity: Effects of Dietary Saturated and Transfatty Acids," *Mediators of Inflammation* 2013 (January 31, 2013): 137579, doi: 10.1155/2013/137579.
12. Fallon and Enig, "Fats," introduction to *Nourishing Traditions* (see chap. 1, n. 40).
13. Sonnenburg and Sonnenburg, "Losing Our Closest Friends," chap. 3 in *Good Gut* (see chap. 1, n. 1).
14. Joseph Mercola, "How Stress Wreaks Havoc on Your Gut – And What to Do about It," *Mercola.com*, April 9, 2012, http://articles.mercola.com/sites/articles/archive/2012/04/09/chronic-stress-gut-effects.aspx.

15. Chris Kresser, "How Stress Wreaks Havoc on Your Gut – and What to Do about It," *ChrisKresser.com*, March 23, 2012, www.chriskresser.com/how-stress-wreaks-havoc-on-your-gut/.
16. "Antimicrobial Resistance," *U.S. Food and Drug Administration*, accessed June 14, 2016, http://www.fda.gov/NewsEvents/PublicHealthFocus/ucm235649.htm.
17. Sonnenburg and Sonnenburg, "Our First Microbiota Inhabitants," chap. 2 in *Good Gut* (see chap. 1, n. 1).
18. Donna Schwenk, "Coming into Culture," chap. 1 in *Cultured Food for Life: How to Make and Serve Delicious Probiotic Foods for Better Health and Wellness* (Carlsbad, CA: Hay House, 2013), Kindle e-book; Josh Axe, "Seven Kefir Benefits and Nutrition Facts," *DrAxe.com*, accessed June 7, 2016, www.draxe.com/kefir-benefits/.
19. Skin Deep, "BHA", *Environmental Working Group*, accessed June 14, 2016, www.ewg.org/skindeep/ingredient/700740/BHA/; Skin Deep, "BHT", *Environmental Working Group*, accessed June 14, 2016, www.ewg.org/skindeep/ingredient/700741/BHT/.
20. Sonnenburg and Sonnenburg, "Premature Birth," chap. 2 in *Good Gut* (see chap. 1, n. 1).
21. Heather Hatfield, "Power Down for Better Sleep," *WebMD*, accessed June 7, 2016, www.webmd.com/sleep-disorders/features/power-down-better-sleep.
22. Arianna Huffington, interview by Marie Forleo, *MarieTV*, YouTube video, March 25, 2014, https://www.youtube.com/watch?v=hpyeggenq2U.
23. Insight Timer, accessed June 16, 2016, https://insighttimer.com/.
24. Dr. Foster and Dr. Smith, "Ingredients in Flee & Tick Control Products," *PetEducation.com*, accessed June 14, 2016, www.peteducation.com/article.cfm?c=2+2111&aid=598.

Chapter 3

1. Anne Marie, "What Is Fermentation?," *Chemistry.About.com*, last modified June 1, 2016, http://chemistry.about.com/od/lecturenoteslab1/f/What-Is-Fermentation.htm.
2. Patrick E. McGovern, Juzhong Zhang, Jigen Tang, Zhiqing Zhang, Gretchen R. Hall, Robert A. Moreau, Alberto Nuñez, Eric D. Butrym, Michael P. Richards, Chen-shan Wang, Guangsheng Cheng, Zhijun Zhao, and Changsui Wang, "Fermented Beverages of Pre- and Proto-historic China," *Proceedings of the National Academy of Sciences of the United States of America* 101, no. 51 (December 8, 2004): 17593-98, doi: 10.1073/pnas.0407921102.
3. Sandor Katz, "The Preservation Benefits of Fermentation, and Their Limits," chap. 2 in *The Art of Fermentation: An In-depth Exploration of Essential Concepts and Processes from around the World* (White River Junction, VT: Chelsea Green Publishing, 2012), Kindle edition.
4. K. M. Cho, R. K. Math, S. M. Islam, W. J. Lim, S. Y. Hong, J. M. Kim, M. G. Yun, J. J. Cho, and H. D. Yun, "Biodegradation of Chlorpyrifos by Lactic Acid Bacteria during Kimchi Fermentation," abstract, *Journal of Agricultural and Food Chemistry* 57, no. 5 (March 11, 2009): 1882-89, doi: 10.1021/jf803649z.
5. Hiromitsu Watanabe, "Beneficial Biological Effects of Miso with Reference to Radiation Injury, Cancer and Hypertension," *Journal of Toxicologic Pathology* 26, no. 2 (July 10 2013): 91-103, doi: 10.1293/tox.26.91.
6. Ibid.
7. Katz, "Detoxification," chap. 2 in *Art of Fermentation*.
8. Katz, "Pre-Digestion," chap. 2, in *Art of Fermentation*.
9. A. S. Hole, I. Rud, S. Grimmer, S. Sigl, J. Narvhus, and S. Sahlstrom, "Improved Bioavailability of Dietary Phenolic Acids in Whole Grain Barley and Oat Groat Following Fermentation with Probiotic *Lactobacillus acidophilus*, *Lactobacillus johsonii*, and *Lactobacillus reuteri*," abstract, *Journal of*

Agricultural and Food Chemistry 60, no. 25 (June 27, 2012): 6369-75, doi: 10.1021/jf300410h.

10. Katz, "Nutritional Enhancement," chap. 2, in *Art of Fermentation*.

11. Katz, "Wild Fermentation Versus Culturing," chap. 3, in *Art of Fermentation*.

12. Stephen Daniells, "Can Fermented Kimchi Alter the Gut Microbiota and Influence Metabolism?," *NITRAIngredients.com*, March 30, 2015, www.nutraingredientsusa.com/research/can-fermented-kimchi-alter-the-gut-microbiota-and-influence-metabolism.

13. Bowe, Patel, and Logan, "Gut-brain-skin Axis," 185-199 (see chap. 1, n. 36).

14. Ö. Özdemir, "Vaious Effects of Different Probiotic Strains in Allergic Disorders: an Update from Laboratory and Clinical Data," *Clinical and Experimental Immunology* 160, no. 3 (June 2010): 295-304, doi: 10.1111/j.1365-2249.2010.04109.x.

15. Mérilie Gagnon, Patricia Savard, Audrey Riviere, Gisèle LaPointe, and Denis Roy, "Bioaccessible Antioxidants in Milk Fermented by *Bifidobacterium longum subsp. longum* Strains," *BioMed Research International* 2015:169381, doi: 10.1155/2015/169381; Soo Im Chung, Catherine W. Rico, and Mi Young Kang, "Comparative Study on the Hypoglycemic and Antioxidative Effects of Fermented Paste (Doenjang) Prepared from Soybean and Brown Rice Mixed with Rice Bran or Red Ginseng Marc in Mice Fed with High Fat Diet," *Nutrients* 6, no. 10 (October 2014): 4610-24, doi: 10.3390/nu6104610; Swee Keong Yeap, Boon Kee Beh, Norlaily Mohd Ali, Hamidah Mohd Yusof, Wang Yong Ho, Soo Peng Koh, Noorjahan Banu Alitheen, and Kamariah Long, "*In Vivo* Antistress and Antioxidant Effects of Fermented and Germinated Mung Bean, *BioMed Research International* 2014:694842, doi: 10.1155/2014/694842.

16. Akimitsu Takagi, Mitsuyoshi Kano, and Chiaki Kaga, "Possibility of Breast Cancer Prevention: Use of Soy Isoflavones and Fermented Soy Beverage Produced Using Probiotics," ed. Sanjay K. Srivastava, *International Journal of Molecular Sciences* 16, no. 5 (May 2015): 10907-20, doi: 10.3390/ijms160510907; K. J. Lee, S. Y. Lee, and G. E. Ji, "Diabetes-Ameliorating Effects

of Fermented Red Ginseng and Causal Effects on Hormonal Interactions: Testing the Hypothesis by Multiple Group Path Analysis," abstract, *Journal of Medicinal Food* 16, no. 5 (May 2013): 383-95, doi: 10.1089/jmf.2012.2583.

17. S. Makino, A. Sato, A. Goto, M. Nakamura, M. Ogawa, Y. Chiba, J. Hemmi, H. Kano, K. Takeda, K. Okumura, and Y. Asami, "Enhance Natural Killer Cell Activation by Exopolysaccharides Derived from Yogurt Fermented with *Lactobacillus delbrueckii ssp. bulgaricus* OLL1073R-1," abstract, *Journal of Daily Science* 99, no. 2 (February 2016): 915-23, doi: 10.3168/jds.2015-10376.

18. N. Borruel, M. Carol, F. Casellas, M. Antolin, F. de Lara, E. Espin, J. Naval, F. Guarner, and J. R. Malagelada, "Increased Mucosal Tumour Necrosis Factor α Production in Crohn's Disease Can Be Downregulated Ex Vivo by Probiotic Bacteria," *Gut* 51, no. 5 (2002): 659-64, doi: 10.1136/gut.51.5.659; "*Lactobacillus casei*", *Probiotic.org*, accessed June 14, 2016, www.probiotic.org/lactobacillus-casei.htm; "*Lactobacillus bulgaricus*," *Probiotic.org*, accessed June 14, 2016, www.probiotic.org/lactobacillus-bulgaricus.htm.

19. Eva M Selhub, Alan C. Logan, and Alison C. Bested, "Fermented Foods, Microbiota, and Mental Health: Ancient Practice Meets Nutritional Psychiatry," *Journal Physiological Anthropology* 33, no. 1 (2014): 2, doi: 10.1186/1880-6805-33-2; "*Lactobacillus helveticus*," *Probiotic.org*, accessed June 14, 2016, http://www.probiotic.org/lactobacillus-helveticus.htm; "*Bididobacterium longum*," *Probiotic.org*, accessed June 14, 2016, http://www.probiotic.org/Bifidobacterium-Longum.htm.

20. Stephanie N. Chilton, Jeremy P. Burton, and Gregor Reid, "Inclusion of Fermented Foods in Food Guides around the World," *Nutrients* 7, no. 1 (January 2015): 390-404, doi: 10.3390/nu7010390.

21. V. K. Shiby and H. N. Mishra, "Fermented Milks and Milk Products as Functional Foods – a Review," abstract, *Critical Reviews in Food Science and Nutrition* 53, no. 5 (2013): 482-96, doi: 10.1080/10408398.2010.547398.

22. A. K. Rai, S. Sanjukta, and K. Jeyaram, "Production of Angiotensin I Converting Enzyme Inhibitory (ACE-I)

Peptides during Milk Fermentation and Their Role in Reducing Hypertension", abstract, *Critical Reviews in Food Science and Nutrition*, October 13, 2015, doi: 10.1080/10408398.2015.1068736; Y.Y. Liu, S. Y. Zeng, Y. L. Leu, and T. Y. Tsai, "Antihypertensive Effect of a Combination of Uracil and Glycerol Derived from Lactobacillus plantarum Strain TWK 10-Fermented Soy Milk," abstract, *Journal of Agricultural and Food Chemistry* 63, no. 33 (August 26, 2015): 7333-7342, doi: 10.1021/acs.jafc.5b01649.

23. Chilton, Burton, and Reid, "Food Guides," 390-404; M. P. St-Onge, E. R. Farnworth, and P. J. Jones, "Consumption of Fermented and Nonfermetned Dairy Products: Effects on Cholesterol Concentrations and Metabolism," abstract, *American Journal of Clinical Nutrition* 71, no. 3 (March 2000): 674-81, www.ncbi.nlm.nih.gov/pubmed/10702159.

24. P. Veiga, N. Pons, A. Agrawal, R. Oozeer, D. Guyonnet, R. Brazeilles, J. M. Faurie, J. E. van Hylckama Vlieg, L. A. Houghton, P. J. Whorwell, S. D. Ehrlich, and S. P. Kennedy, "Changes of the Human Gut Microbiome Induced by a Fermented Milk Product", abstract, *Scientific Reports* 4 (September 11, 2014): 6328, doi: 10.1038/srep06328; L. Moussa, V. Bézirard, C. Salvador-Cartier, V Bacquié, E. Houdeau, and V. Théodorou, "A New Soy Germ Fermented Ingredient Displays Estrogenic and Protease Inhibitor Activities Able to Prevent Irritable Bowel Syndrome-like Symptoms in Stressed Female Rats," abstract, *Clinical Nutrition* 32, no. 1 (February 2013): 51-58, doi: 10.1016/j.clnu.2012.05.021.

25. Joonki Kim, Sung Hun Kim, Deuk-Sik Lee, Dong-Jin Lee, Soo-Lyun Kim, Sungkwon Chung, and Hyun Ok Yang, "Effects of Fermented Ginseng on Memory Impairment and β-amyloid Reduction in Alzheimer's Disease Experimental Models," *Journal of Ginseng Research* 37, no. 1 (March 2013): 100-107, doi: 10.5142/jgr.2013.37.100.

26. Mariángeles Noto Llana, Sebastián Hernán Sarnacki, María del Rosario Aya Castañeda, María Isabel Bernal, Mónica Nancy Giacomodonato, and María Cristina Cerquetti, "Consumption

of *Lactobacillus casei* Fermented Milk Prevents Salmonella Reactive Arthritis by Mondulating IL-23/IL-17 Expression," ed. Josep Bassaganya-Riera, *PLos ONE* 8, no. 12 (2013): e82588, doi: 10.1371/journal.pone.0082588; Manas Ranjan Swain, Marimuthu Anandharaj, Ramesh Chandra Ray, and Rizwana Parveen Rani, "Fermented Fruits and Vegetables of Asia: A Potential Source of Probiotics," *Biotechnology Research International* 2014:250424, doi: 10.1155/2014/250424.

27. Sein Lee, Jong-Eun Kim, Sujin Suk, Oh Wook Kwon, Gaeun Park, Tae-gyu Lim, Sang Gwon Seo, Jong Rhan Kim, Dae Eung Kim, Miyeong Lee, Dae Kyun Chung, Jong Eun Jeon, Dong Woon Cho, Byung Serk Hurh, Sun Yeou Kim, and Ki Won Lee, "A Fermented Barley and Soybean Formula Enhances Skin Hydration", *Journal of Clinical Biochemistry and Nutrition* 57, no. 2 (September 2015): 156-63, doi: 10.3164/jcbn.15-43.

28. Doctor Natasha's personal website, accessed June 14, 2016, http://www.doctor-natasha.com.

29. Katz, "Fermenting Vegetables (and Some Fruits Too)", chap. 5 in *Art of Fermentation*.

30. Katz, "Lactic Acid Bacteria," chap. 5 in *Art of Fermentation*.

31. Katz, "Which Vegetables Can Be Fermented?," chap. 5 in *Art of Fermentation*.

32. Ibid.

Chapter 4

1. Ben Kim, "The Truth about Alkalizing Your Blood," *DrBenKim.com*, July 4, 2014, www.drbenkim.com/ph-body-blood-foods-acid-alkaline.htm.

2. "Alkalosis," *U.S. National Library of Medicine, National Institutes of Health*, last modified November 1, 2015, https://www.nlm.nih.gov/medlineplus/ency/article/001183.htm.

3. Kim, "Alkalizing Your Blood."
4. "Is Peanut Butter Healthy? Yes, Says the Harvard Heart Letter," *Harvard Health Publications*, accessed June 14, 2016, http://www.health.harvard.edu/press_releases/is-peanut-butter-healthy.
5. Katz, "Natto," chap. 11 in *Art of Fermentation* (see chap. 3, n. 3); Phil Domenico, "The Acid-Alkaline Food Guide: Interview with the Author," *Natural News*, July 24, 2008, www.naturalnews.com/023694_food_foods_health.html; "Acid-forming & Alkaline-forming Foods", *Edgar Cayce's Association for Research and Enlightenment*, accessed June 16, 2016, http://www.edgarcayce.org/are/holistic_health/data/thdiet3.html.
6. Joseph Mercola, "15 Natural Remedies for the Treatment of Acid Reflux and Ulcers," *Mercola.com*, April 28, 2014, articles.mercola.com/sites/articles/archive/2014/04/28/acid-reflux-ulcer-treatment.aspx; Kathryne Pirtle, "A Precursor to Chronic Illness," *Modern Diseases (blog)*, *The Weston A. Price Foundation*, June 25, 2010, www.westonaprice.org/modern-diseases/acid-reflux-a-red-flag/.
7. Pirtle, "Chronic Illness."
8. Andrew C. Dukowicz, Brian E. Lacy, and Gary M. Levine, "Small Intestinal Bacterial Overgrowth: A Comprehensive Review," *Gastroenterology & Hepatology* 3, no. 2 (February 2007): 112-22: http://www.ncbi.nlm.nih.gov/pmc/articles/PMC3099351/pdf/GH-03-112.pdf, PDF document; Jen Broyles, "Learn from the Best – Dr. Allison Siebecker Talks SIBO," *JenBroyles.com*, May 12, 2015, www.jenbroyles.com/dr-allison-siebecker-talks-sibo/.
9. Dukowicz, Lacy, and Levine, "Small Intestine Bacterial Overgrowth," 112-22.
10. Joseph Mercola, "Pure Water: This Simple Drink Improved Stomach Acid in Just One Minute...," *Mercola.com*, July 8, 2011, articles.mercola.com/sites/articles/archive/2011/07/08/water-works-better-than-ulcer-pills-to-decrease-stomach-acid.aspx.
11. Donna Swchenk, "Fibromyalgia," chap. 4 in *Cultured Food for Health* (Carlsbad, CA: Hay House, 2015), Kindle ebook. .

12. Sara Tomm, "Complications from Too Much Sodium," *Healthy Eating* (blog), *SFGate.com*, accessed June 17, 2016, http://healthyeating.sfgate.com/complications-much-sodium-5874.html.

13. Queen's University, "New Link Between Wine, Fermented Food and Cancer," *Science Daily*, March 8, 2007, www.sciencedaily.com/releases/2007/03/070307152917.htm.

14. Barbara Demick, "Korean's Kimchi Adulation, with a Side of Skepticism," *Los Angeles Times*, May 21, 2006, articles.latimes.com/2006/may/21/world/fg-kimchi21.

15. Dr. Mark Hyman's personal website, accessed June 17, 2016, http://drhyman.com/; Dr. Joseph Mercola's personal website, accessed June 17, 2016, http://www.mercola.com/; Dr. Andrew Weil's personal website, accessed June 17, 2016, http://www.drweil.com/.

16. Michael Han, December 3, 2013, comment on "Does Kimchi Cause Stomach Cancer?," *Quora*, accessed June 17, 2016, https://www.quora.com/Does-kimchi-cause-stomach-cancer.

17. Chris Kresser, "The Nitrate and Nitrite Myth: Another Reason Not to Fear Bacon," *ChrisKresser.com*, October 5, 2012, chriskresser.com/the-nitrate-and-nitrite-myth-another-reason-not-to-fear-bacon/.

18. Kris Gunnars, "Are Nitrates and Nitrites in Foods Harmful?," *Authority Nutrition*, December 2015, authoritynutrition.com/are-nitrates-and-nitrites-harmful.

19. "Vital Signs: Food Categories Contributing the Most to Sodium Consumption – United States, 2007 – 2008," *Morbidity and Mortality Weekly Report*, Centers for Disease Control and Prevention 61, no. 5 (February 10, 2012): 92-98, www.cdc.gov/mmwr/preview/mmwrhtml/mm6105a3.htm?s_cid=mm6105a3_w.

20. "Lowering Salt in Your Diet," *Consumer Updates, U.S. Food and Drug Administration*, accessed June 14, 2016, www.fda.gov/ForConsumers/ConsumerUpdates/ucm181577.htm.

21. Information about salt intake, *Centers for Disease Control and Prevention*, accessed June 14, 2016, http://www.cdc.gov/salt/.

22. National Center for Chronic Disease Prevention and Health Promotion, "Where's the Sodium?," *Centers for Disease Control*

and *Prevention*, last modified February 2, 2016, www.cdc.gov/vitalsigns/Sodium/index.html.

23. "Added Sugars: Don't Get Sabotaged by Sweeteners," *Healthy Lifestyle, Mayo Clinic*, accessed June 14, 2016, www.mayoclinic.org/healthy-lifestyle/nutrition-and-healthy-eating/in-depth/added-sugar/art-20045328; Julie Corliss, "Eating Too Much Added Sugar Increases the Risk of Dying with Heart Disease," *Harvard Health Blog, Harvard Health Publications*, February 6, 2014, www.health.harvard.edu/blog/eating-too-much-added-sugar-increases-the-risk-of-dying-with-heart-disease-201402067021.

24. Sonnenburg and Sonnenburg, "Eating for Your Microbes," chap. 9 in *Good Gut* (see chap. 1, n. 1).

25. "Sodium Benzoate," *Listing of Specific Substances Affirmed as GRAS, Direct Food Substances Affirmed as Generally Recognized As Safe, Food and Drug Administration*, last modified August 21, 2015, https://www.accessdata.fda.gov/scripts/cdrh/cfdocs/cfcfr/CFRSearch.cfm?fr=184.1733.

26. Martin Downs, "The Truth about Seven Common Food Additives," *WebMD*, December 17, 2008, www.webmd.com/diet/the-truth-about-seven-common-food-additives?page=1.

27. Jeannine Stein, "Skinnygirl Margarita Pulled: What Is Sodium Benzoate; Is It Bad?," *Los Angeles Times*, September 8, 2011, articles.latimes.com/2011/sep/08/news/la-heb-skinnygirl-margarita-sodium-benzoate-20110908.

28. "Polysorbate 80," *Multipurpose Additives, Food Additives Permitted for Direct Addition to Food for Human Consumption, Food and Drug Administration*, last modified August 21, 2015, http://www.accessdata.fda.gov/scripts/cdrh/cfdocs/cfcfr/cfrsearch.cfm?fr=172.840; E. A. Coors, H. Seybold, H. F. Merk, and V. Mahler, "Polysorbate 80 in Medical Products and Nonimmunologic Anaphylactoid Reactions," abstract, *Annuals of Allergy, Asthma, and Immunology* 95, no. 6 (December 2005): 593-99, www.ncbi.nlm.nih.gov/pubmed/16400901.

29. Anne Marie, "Is Alum Safe to Eat or Use?," *Chemistry.About.com*, last modified February 16, 2016, chemistry.about.com/od/foodchemistryfaqs/f/Is-Alum-Safe.htm.

30. Katz, "Live Bacterial Cultures," chap. 2 in *Art of Fermentation* (see chap. 3, n. 3).

31. Joseph Mercola, "Pregnancy Alters Resident Gut Microbes," *Mercola.com*, August 23, 2012, articles.mercola.com/sites/articles/archive/2012/08/23/trimester-pregnancy-affects-baby-health.aspx.

32. Joseph Mercola, "Why Your Gut Flora Powerfully Dictates Whether You're Healthy or Sick," *Mercola.com*, March 18, 2012, articles.mercola.com/sites/articles/archive/2012/08/23/trimester-pregnancy-affects-baby-health.aspx.

33. Jill Nienhiser, "Diet for Pregnant and Nursing Mothers," *Children's Health (blog)*, *The Weston A. Price Foundation*, January 10, 2004, www.westonaprice.org/childrens-health/diet-for-pregnant-and-nursing-mothers/.

34. Mercola, "Resident Gut Microbes."

35. Carol T. Culhane, "Everything You Want to Know about Gluten," *Institute of Food Technologists*, accessed June 15, 2016, http://www.ift.org/knowledge-center/learn-about-food-science/food-facts/gluten.aspx.

36. "Gluten-free Fermented Foods," *Cultures for Health*, accessed June 15, 2016, www.culturesforhealth.com/learn/general/gluten-free-fermented-foods/.

37. Katz, "Live Bacterial Cultures," chap. 2 in *Art of Fermentation* (see chap. 3, n. 3).

38. Joseph Mercola, "Fermented Foods Contain 100 TIMES More Probiotics than a Supplement," *Mercola.com*, May 12, 2012, articles.mercola.com/sites/articles/archive/2012/05/12/dr-campbell-mcbride-on-gaps.aspx.

39. Sonnenburg and Sonnenburg, "Not Just a Gut Effect," chap. 4 in *Good Gut* (see chap. 1, n, 1).

40. Sonnenburg and Sonnenburg, "What's In a Name?," chap. 4 in *Good Gut* (see chap. 1, n, 1).

41. "NIH Human Microbiome Project Defines Normal Bacterial Makeup of the Body: Genome Sequencing Creates First Reference Data for Microbes Living with Healthy Adults," *National Institutes of Health*, June 13, 2012, www.nih.gov/news-

events/news-releases/nih-human-microbiome-project-defines-normal-bacterial-makeup-body/

42. Amy Myers, "Everything You Need to Know about Histamine Intolerance," *Mind Body Green*, October 3, 2013, www.mindbodygreen.com/0-11175/everything-you-need-to-know-about-histamine-intolerance.html.

43. Chris Kresser, "Headaches, Hives, and Heartburn: Could Histamine Be the Cause," *ChrisKresser.com*, January 25, 2013, chriskresser.com/headaches-hives-and-heartburn-could-histamine-be-the-cause/.

44. Myers, "Histamine Intolerance."

45. "Histamine Hack: How to Safely Eat Fermented Foods," *Body Ecology*, accessed June 17, 2016, bodyecology.com/articles/histamine-hack-how-to-safely-eat-fermented-foods.

Chapter 5

1. "Sustainability & Safety," *Ziploc*, accessed June 17, 2016, https://ziploc.com/en/sustainability-and-safety.

2. "Chemical Compatibility Guide," *Sevier Lab, Cornell University*, accessed June 15, 2016, sevierlab.vet.cornell.edu/resources/Chemical-Resistance-Chart-Detail.pdf.

3. Ibid.

4. Kirsten K. Shockey and Christopher Shockey, "Temperature and Light," chap. 2 in *Fermented Vegetables: Creative Recipes for Fermenting 64 Vegetables & Herbs in Krauts, Kimchis, Brined Pickles, Relishes & Pastes* (North Adams, MA: Storey Publishing, 2014), Kindle ebook.

5. Caroline Barringer, interview by Joseph Mercola, *Mercola.com Official YouTube Channel*, YouTube video, December 21, 2011, https://www.youtube.com/watch?v=hy87TnyNCPk.

6. Shockey and Shockey, "Temperature and Light," chap. 2 in *Fermented*.

7. Shockey and Shockey, "Know When It's Done," chap. 4 in *Fermented*.
8. Shockey and Shockey, "How Salt Works," chap. 2 in *Fermented*.
9. Ibid.
10. Cookson Beecher, "Fermenting Veggies at Home: Follow Food Safety ABCs," *Food Safety News*, March 11, 2014, www.food safetynews.com/2014/03/fermenting-veggies-at-home-follow-food-safety-abcs/#.V1cSCJErJaS.

Chapter 6

1. Katz, "Fermenting Vegetables (and Some Fruits too)", chap. 5 in *Art of Fermentation* (see chap. 3, n. 3).
2. Shockey and Shockey, "The Fermentista's Mantra: The Path to Success," part 2 in *Fermented Vegetables* (see chap. 5, n. 4).
3. Alex Lewin, "Basic Sauerkraut," chap. 3 in *Real Food Fermentation: Preserving Whole Fresh Food with Live Cultures in Your Home Kitchen* (Beverly, MA: Quarry Books, 2012), 62.
4. Barringer, *Mercola.com* (see chap. 5, n. 5).
5. These are tips from Kirsten Shockey from *Fermented Vegetables* when I had an interview with her.

Chapter 7

1. Barringer, *Mercola.com* (see chap. 5, n. 5)
2. Fallon and Enig, "Fermented Vegetables & Fruits," *Nourishing Traditions* (see chap. 1, n. 40).
3. Barringer, *Mercola.com* (see chap. 5, n. 5).
4. Fallon and Enig, "Fermented Vegetables & Fruits," *Nourishing Traditions* (see chap. 1, n. 40).

5. Swchenk, *Cultured Food for Life* (see chap. 2, n. 18); Swchenk, *Cultured Food for Health* (see chap. 4, n. 11).

6. Shockey and Shockey, "Dessert," chap. 14 in *Fermented Vegetables* (see chap. 5, n. 4).

7. Michael Pollan, introduction to *In Defense of Foods* (New York: The Penguin Press, 2008).

8. Josepher Mercola, "Surprising Health Benefits of Vegetables," *Mercola.com*, September 8, 2014, articles.mercola.com/sites/articles/archive/2014/09/08/vegetable-health-benefits.aspx.

9. Lydia A. Bazzano, Tricia Y. Li, Kamudi J. Joshipura, and Frank B. Hu, "Intake of Fruit, Vegetables, and Fruit Juices and Risk of Diabetes in Women," *Diabetes Care* 31, no. 7 (July 2008): 1311-17, doi: 10.2337/dc08-0080.

10. Mercola, "Benefits of Vegetables."

11. Katz, "Live Bacterial Cultures," chap. 2 in *Art of Fermentation* (see chap. 3, n. 3).

12. Fallon and Enig, "Fermented Vegetables & Fruits," *Nourishing Traditions* (see chap. 1, n. 40).

13. Schwenk, "Basic Cultured Vegetables," chap. 6 in *Cultured Food for Health* (see chap. 4, n. 11).

Chapter 8

1. Ivan D. Jones and John L. Etchells, "Nutritive Value of Brined and Fermented Vegetables," *American Journal of Public Health and the Nations Health* 34, no. 7 (July 1944): 717, www.ncbi.nlm.nih.gov/pmc/articles/PMC1625074/?page=7.

Chapter 9

1. Katz, "Selective Environments," chap. 3 in *Art of Fermentation* (see chap. 3, n. 3).
2. Boston Ferments, accessed June 15, 2016, www.bostonferments.com.

Chapter 12

1. Lewin, "Basic Sauerkraut," chap. 3 in *Real Food Fermentation*, 61-64 (see chap.6, n. 3).
2. Barringer, *Mercola.com* (see chap. 5, n. 5).
3. Shockey and Shockey, "Lemon Spinach," part 3 in *Fermented Vegetables* (see chap. 5, n. 4).
4. Shockey and Shockey, "Fermented Eggplant," part 3 in *Fermented Vegetables* (see chap. 5, n. 4).
5. Shockey and Shockey, "Leek Paste," part 3 in *Fermented Vegetables* (see chap. 5, n. 4).
6. Shockey and Shockey, "Whole-Leaf Ferments," chap. 5 in *Fermented Vegetables* (see chap. 5, n. 4).
7. Katz, "Fermenting Vegetables (and Some Fruits too), chap. 5 in *Art of Fermentation* (see chap. 3, n. 3).
8. Lewin, "Cucumber Pickles," chap. 4 in *Real Food Fermentation*, 83-85 (see chap.6, n. 3).
9. Karen Solomon, "Tianjin Preserved Vegetable," *Asian Pickles: Sweet, Sour, Salty, Cured, and Fermented Preserves* (Berkeley, CA: Ten Speed Press, 2014), 82-85
10. Shockey and Shockey, "Basic Steps for Making Kimchi," chap. 7 in *Fermented Vegetables* (see chap. 5, n. 4).
11. Sandor Katz, "Radish and Root Kimchi," chap. 5 in *Wild Fermentation: The Flavor, Nutrition, and Craft of Live-Culture Foods* (White River Junction, VT: Chelsea Green Publishing Company, 2003), Kindle edition.

12. Shockey and Shockey, "Asparagus Kimchi," part 3 in *Fermented Vegetables* (see chap. 5, n. 4).

Chapter 13

1. Fallon and Enig, "Carbohydrates," introduction to *Nourishing Traditions* (see chap. 1, n. 40); Fallon and Enig, "Grains & Legumes" in *Nourishing Traditions* (see chap. 1, n. 40).

ACKNOWLEDGEMENTS

Initially, I wrote this book with the sole intention to find out a detailed step-by-step process on how to make fermented vegetables and how to eat and preserve them properly. This more-than-a-year-long endeavor has turned out to be more rewarding than I thought.

From my research progress, I switched my mindset from "what I wanted to know" to "how I could use this book to help other address their concerns and solve their problems". The turning point was when I conducted a survey among 100 people. I thank those who participated in my survey and shared concerns with me related to making and consuming fermented foods for helping me find meanings for writing this book to make it truly be of service.

In my research process, my knowledge and interest grew as I attended workshops, potluck parties, and other gatherings hosted by Boston Ferments (BF). I thank everyone who volunteers for this organization for making every event possible.

My trip to Boston Fermentation Festival hosted by BF in 2015 was exceptionally helpful to my research. I'd like to thank organizers for making such a fun, educational, and interactive event. This event allowed me to taste assorted fermented foods like sourdough topped with cranberries and pickled root vegetables, to learn more about techniques of cucumber pickling and salting, and to connect with scientists and authors who helped address my concerns and other questions I received.

Particularly, I am, and will be, forever grateful for the experts who agreed to spend hour-long, if not longer, conversations with me answering my questions via Skype and phone. They are: Sandor Katz (author of *The Art of Fermentation* and *Wild Fermentation*), Donna Schwenk (author of *Cultured Food for Life* and *Cultured Food for Health*), Kirsten Shockey (co-author of *Fermented Vegetables*), and Frederick Breidt (microbiologist working for USDA specialized in vegetable fermentation).

I would also like to thank Yi Luo, a Traditional Chinese Medicine (TCM) practitioner based in Massachusetts, who helped me learn more about TCM and how this philosophy is applied to our health. Despite her busy schedule and not having time for an interview with me, she asked me to type out a list of questions, so that she could reply to me via texts when she found time.

I also want to thank all of my friends who show interest in and support to what I am doing, my accountability buddy Blaine who gave a tremendous amount of encouragement and always kept me on track at times when I deviated from my writing goals or procrastinated.

In addition, I thank Tom Morkes, founder of Publishers' Empire, a self-publishing program that provides thorough and insightful information on how to write, package, and market my own book. This platform has also helped me refine my book subjects so that they focus on truly helping people solve problems. This whole writing process has not only sharpened my writing skills, but also allowed me to grow and constantly challenge myself to break out of boundaries.

I thank Tom for creating such an awesome platform that teaches me how to share my message while seeking personal growth. I also thank the community I am with inside Publishers' Empire; and I really appreciate this environment of mutual support, from which I have learned how to use writing as a means to serve people that need help.

I would also like to thank Cyrus Kirkpatrick, my amazing editor for helping me craft my message in a most effective way. I see writing this book as a process of creating a statue out of a piece of wood; Cyrus has helped me chip away what the statue is not, so that only what belongs to it gets to stay. He provides independent editing service; if you need his help, I encourage you to look into his website at www.cyruskirkpatrick.com/partnership.

Additionally, I thank my book cover design Lara Iakovenko for her exceptional work and service. She patiently worked on the prototype to make sure that it reflects the key messages this book is trying to deliver. It was such a pleasure working with her. You can learn more about Lara via her site at https://www.behance.net/jiskra.

Of course, I would like to thank my parents for giving my life, so that I am able to, and can continue to, find and fulfill my purpose in this current lifetime. I thank my parents for giving me endless love, caring, and support along the way.

Last but not least, I would like to thank my partner Christopher. He was the guinea pig for most of my vegetable fermentation experiments and managed to put up with a messy kitchen and a countertop covered by Mason jars, lids, stainless steel bowls, vegetables leaves, and garlic smells. I thank him for tolerating every crazy home experiment I have made and for believing in whatever I do.